THE
FRUITS
OF
WAR

THE FRUITS OF WAR

How Military Conflict Accelerates Technology

MICHAEL WHITE

SIMON &
SCHUSTER

London · New York · Sydney · Toronto · Dublin

A VIACOM COMPANY

First published in Great Britain by Simon & Schuster UK Ltd, 2005
A Viacom Company

1 3 5 7 9 10 8 6 4 2

Simon & Schuster UK Ltd
Africa House
64–78 Kingsway
London WC2B 6AH

www.simonsays.co.uk

Simon & Schuster Australia
Sydney

A CIP catalogue record for this book is available
from the British Library.

Hardback ISBN 0-7432-2024-2
EAN 9780743220248

Trade paperback ISBN 0-7432-8529-8
EAN 9780743285292

Typeset in Melior by M Rules
Printed and bound in Great Britain by The Bath Press, Bath

For Lisa again, but with more love

Contents

CONTENTS

I will ignore all ideas for new works and engines of war, the invention of which has reached its limits and for whose improvement I see no further hope.

Julius Frontinus, chief military engineer to
the Emperor Vespasian, *c.* AD 70

Introduction: Good From Bad

Look around you.

Where are you? On a train, perhaps? Sitting on an airliner at thirty thousand feet? Or are you seated comfortably on a sofa in your living room sipping a glass of wine? Wherever you may be, you are surrounded by the accoutrements of the twenty-first century, cocooned by technology.

A few feet away from you are machines, cars, TVs, aircraft guidance systems; satellites whirl overhead. In your pockets you may have a mobile phone and a palmtop, in your stomach an antibiotic tablet dissolving slowly. All these things have something in common, all are linked, connected by the invisible thread that created them. Each has evolved from a common source.

So have you.

You are a product of the techno age, a product of the many things that have been forged from steel and silicon, sweat and blood. You are a child of technology, a child of war.

'War,' the medical historian Professor Roy Porter once wrote, '. . . is often good for medicine. It gives the medical profession ample opportunities to develop its skills and hone its practices. It can also create a post-war mood eager to beat swords into scalpels.' But while poignant and accurate, this statement could be broadened enormously to encompass many other disciplines and a rich variety of technologies.

The horror and insanity of military conflict is ever with us; it is a constant. Our aggression is linked inextricably, many psychologists suppose, with human creative energy. It is the evil twin born from the same seed as our creative urge, the urge that motivates us to learn, to explore. But even our need to wage war and our enthusiasm for conflict that history demonstrates so dramatically also produces much that is positive; for, through the ages, the many sufferings on the battlefield have serendipitously given rise to much that benefits the civilian. As the American historian William H. McNeill has written, 'Anyone looking at the equipment installed in a modern house will readily recognise how much we in the late twentieth century are indebted to industrial changes pioneered in near-panic circumstances when more and more shells, gun-powder, and machine guns suddenly became the price of survival as a sovereign state.'[1]

The acceleration of technology through the impetus of war can happen in different ways. A discovery may stem from a single, inspired idea prompted by a random occurrence in the turmoil of battle. The development of this idea may then follow a torturous path. An alternative route begins with a piece of open research conducted in universities or commercial centres which attracts the attention of the military who then provide the resources for an accelerated development programme. The end result is then used by the military, only later to find its way into civilian life.

A contemporary example of this last is a piece of research into the structure of the eye currently underway at Moorfields Eye Hospital in London. For some time now, high-street chemists have offered a technique called LASIK (Laser-Assisted In Situ Keratomileusis), a method by which, using a specially designed laser, the ophthalmologist can remodel the surface of the cornea to counteract simple aberrations in vision. The team at Moorfields Hospital is working with a far more advanced version of this which enables them to skim away layers of the cornea just a few molecules thick. The team is also investigating the potential of implants that boost the 'normal' visual range. These techniques may become useful not only to repair faulty vision (as LASIK does), but to produce what has been dubbed 'super vision'.

Patients who have their corneas carefully remodelled may possibly see frequencies of the electromagnetic spectrum far beyond the normal range and have a clarity of vision far better than anyone has yet had.

This new work has obvious potential for military applications and the Ministry of Defence in Britain has been quick to realise it and to pour money into the Moorfields project. In return they will expect to have first pickings from developments and the right to pass this information on to other teams to adapt the technique for military use. Later, a different version of the work will evolve into civilian research programmes with the promise of more advanced versions of LASIK surgery available to all.

Such feedback systems between military and civilian organisations provide the backbone for a vast array of technological developments, what have been dubbed 'dual-use technologies'. In the United States this symbiosis is dominated by a single organisation, the Defense Advanced Research Projects Agency (DARPA), established in 1958, at the height of the Cold War, as a direct response to the launching of Sputnik I by the Soviet Union. DARPA was a militarised version of the civilian-controlled Office of Scientific Research and Development (OSRD), established by the computer visionary Vannevar Bush. Its remit is to develop embryonic technologies that could be made useful to the Defense Department in Washington, and in meeting this goal it has been remarkably successful. In Britain, DARPA's equivalent is the MOD's Defence Evaluation and Research Agency (DERA).

If we pause to look at DARPA's achievements, it soon becomes apparent that this organisation, now operating with an annual budget of $3 billion, has been responsible for a stunning array of technological advances since its creation less than half a century ago. Their most famous contribution is the internet, designed specifically for the Pentagon which wanted a communication network that could survive a nuclear war. But DARPA has provided society with a great deal more. DARPA research has advanced our understanding of such diverse disciplines as behavioural sciences, advanced materials technology, surveillance technology, the multitudinous applications of radar, infrared sensing, X-ray/gamma ray detection systems, satellite communications, high-energy laser technology, sub-micron electronic technology (including the latest microprocessors) and advanced fuel systems for land, sea and air vehicles. Today, DARPA is developing the technology we will all use in the coming decades, a vast spectrum of ideas including new forms of computers (especially bio-cybernetic systems), cybernetic sensors for use in android devices and high-resolution plasma screens for jet fighters (which will also improve domestic TV systems). DARPA is also keen to develop faster and more efficient manufacturing methods for military equipment that could be applied some day to Third World civilian industrial systems that are now on DARPA's drawing boards and which will soon be available commercially, ranging from effective computer voice-recognition software to nanotechnology (the science that may soon enable microscopically small robots to enter the body and operate on damaged or diseased tissue).

A powerful way of accelerating this stream of development is to nurture the cross-fertilisation of disciplines. During the 1970s it was realised that some of the greatest achievements in the history of science have come from a melding of disparate disciplines. Within the domain of theoretical science, Darwin achieved his theory of evolution by blending his knowledge of biology, geology and sociology. Today, Stephen Hawking and other physicists seek what they call a grand unified theory using a heady combination of mathematics, cosmology and particle physics. So, too, the practical scientist and the engineer achieve their greatest insights by combining information from cutting-edge technology to develop an extremely versatile and powerful gestalt.

Research led by a California-based company called Telesensory with funding through DARPA has recently begun to merge virtual-reality (VR) research with medical research. The result is a prototype device that enables an air force pilot to 'see' far more than he would with the naked eye. Sensors on the exterior of the plane feed information to a specially designed headset. From there, signals are relayed directly to the retina of the pilot's eye and his or her brain. An aircraft fitted with such a device needs no windows. Meanwhile, researchers in Japan, the US and Europe are working on ways to use this VR technology to restore sight to the blind.

There is, of course, nothing new about combining military money with the mind of the scientist. Perhaps the earliest recorded example of military patronage (and, indeed, the first example of an arms race) comes from the third century BC when, after entering a competition funded by his government, Archimedes designed a revolutionary new catapult for their army. This device was used to devastating effect at the Battle of Syracuse in 265 BC. Meanwhile, far from Greece, Chinese alchemists of the Han Dynasty labouring to satisfy the demands of their military leaders stumbled upon gunpowder.

For many of the great thinkers of history, military patronage was a vital means of making a living. Brunelleschi, Alberti and, later, Galileo all relied for their living upon the warmongering propensities of their wealthy masters. Indeed, for many years, Leonardo da Vinci (a pacifist) was employed as a military engineer by both the Duke of Milan, Ludovico Sforza, and the infamously cruel Cesare Borgia. In his introductory letter to Sforza, Leonardo declared: 'Most Illustrious Lord, having by now sufficiently considered the experience of those men who claim to be skilled inventors of machines of war, and having realised that the said machines in no way differ from the commonly employed, I shall endeavour, without prejudice to anyone else, to reveal my secrets to Your Excellency, for whom I offer to execute, at your convenience, all the items briefly noted below.' He then went on to describe his designs for siege engines, new forms of gunpowder, artillery and transport devices, along with mortars: '. . . that are very practical and easy to transport, with which I can project stones so that they seem to be raining down, and their smoke will plunge the enemy into terror, to his great hurt and confusion'.[2]

The patronage of leaders whose main interest lay in forging a military advantage over their neighbours also led to the earliest maps and the science of cartography. Indeed, it is no exaggeration to say that an awareness of strategic advantage was every bit as powerful as the allure of gold, and it encouraged the Spanish, the French, the Portuguese and the English to finance expeditions to the furthest points of the globe.

Much later, as the nineteenth century drew to a close, another pioneering soul, Guglielmo Marconi, was forced into accepting military patronage when his revolutionary ideas fell on deaf ears. In 1894, the Italian government refused to fund his research into the transmission and reception of radio waves, but, two years later, the British government, then actively looking for better ways of ship-to-ship communications, financed Marconi and the first radio was patented. By the turn of the century, British dreadnoughts were equipped with the new Marconi devices. From its military application, radio spread across the world and into everyday life. By the 1930s, most homes in developed countries possessed a radio.

The creation of radar followed the same pattern. Almost at the same time that Marconi was developing radio, a German, Christian Hulsmeyer, became fascinated with the work of his countryman Heinrich Hertz, who, twenty years earlier, had discovered that electromagnetic radiation of certain wavelengths was reflected by solid objects. Hulsmeyer built a machine that could measure this effect and during the 1920s two Americans, Breit and Tuve, showed that pulses could be bounced off the ionosphere. None of these developments sparked any interest with those holding the purse strings of government until, in 1939, as war in Europe loomed large, the British government started listening to the claims of a young researcher named Robert Watson-Watt who designed the first working radar system just in time to help the RAF defeat the Luftwaffe in the Battle of Britain.

Other forms of military patronage have saved entire nations from defeat, albeit, in some cases, only temporarily. Most famously, in 1915 the German government realised the country was rapidly using up its reserves of ammonia, essential to the manufacture of high explosives. At that time the only source of the chemical was Chilean nitrates and these could not be obtained because of a British naval blockade; and so, in desperation, the Germans turned to science for help.

A few years before the war, Fritz Haber, the most famous chemist of his time, had devised a way of synthesising ammonia in the laboratory. The problem for Haber and the German war effort was that his process only produced very small quantities of ammonia. With surprising foresight, the German government swiftly concluded that the outcome of the war might depend upon the results of this research and they diverted funding and resources into developing Haber's system. By late 1916, when reserves of nitrates were almost exhausted, Haber had managed to increase the efficiency of his method and began synthesising large quantities of ammonia gas. By the end of the war, all German explosives were being made using his technique. So dramatic was this work that its success, some claim, prolonged the First World War by at least a year.

Haber's work certainly led to the untimely death of many hundreds of thousands of men, but the non-military importance of his method cannot be underestimated, for now ammonia is not only used to make high explosives but it is also an essential component of fertilisers, powerful detergents and modern refrigerants. Today, ammonia is still synthesised using Haber's method and the fertilisers made with it help grow the crops that feed hundreds of millions of people who would otherwise starve. Furthermore, ammonia-based compounds are used extensively as disinfectants all over the world and their use as a refrigerant is enormously valuable to industry.

Yet Haber's story is neither the most profound nor the most dramatic example of military patronage in the history of our times. This honour goes to two projects from the middle of the twentieth century, one sprung from the global carnage of the Second World War, the other from the global tension of Cold War politics during the 1950s.

The making of the atomic bomb and the space race were both enormously expensive enterprises and the Manhattan Project led to the death of hundreds of thousands of Japanese civilians, but it is no exaggeration to say that, collectively, the technological gains from these two endeavours have changed the world. The knowledge accumulated by the team of scientists working to build the first atomic bomb at Los Alamos, New Mexico, took atomic science from a set of theoretical ideas formulated at the beginning of the twentieth century and turned them into practical science. This science has given us atomic power, lasers, fibre optics, mini-computers – a cornucopia of developments that underpin twenty-first century life.

From the space race have come communication satellites, global communication networks, weather monitoring systems, resource-seeking satellites, medical breakthroughs only possible from zero-gravity studies, new drugs, new smart materials, digital technology, miniaturisation; again, the list is vast and impressive.

Sir Solly Zuckerman who has written at length on science and war has pointed out: 'The volume of national resources which is devoted to the applications of science for military purposes is usually enormous relative to that which governments set aside in the civil sphere . . . The limits to which this science is forced in the military sphere are in the end set by the patience and the persuasive powers – and persuasive in several directions – of the men who have the responsibility of seeing that the presumed needs of national security are met.'[3] Indeed, it is clear that often scientific endeavours require such enormous resources that they would never begin if it were not for the influence of military imperative. The two obvious examples of this are the development of applied nuclear science and space travel.

The influence of government and the military is clear in cases such as the Manhattan Project and the space programmes of the United States and Russia, but influence manifests itself in many forms. The Manhattan Project was the first example of institutionalised Big Science. Two other means by which the military needs and the *perceived* military needs of a nation are met are more common, and because of their number and pervasiveness they have been just as important as the multi-billion dollar mega-projects.

The first of these efforts is the enormous military component of university funding. Since the 1940s, a substantial proportion of the billions of dollars of university funding in the West has emanated from the military. The other is the utilisation of small-industry inventiveness. A loop is created in which a novel idea derives from the civilian world – a solo inventor, perhaps, or a small industrial-based team; this idea fails to attract investment in the commercial sector and remains in limbo, half-formed, until military funding (most often in a time of great need) leads to proper development of the idea into a practical device or technique. After the military have had first option on the fruits of this advance, the innovation is made available to the public. This colonising of civilian science by the military means that military application of a design, a device or a technique is usually ten years ahead of any mainstream commercial application.

It is important to note that scientific progress does not derive solely from military impetus, nor only from laboratories funded by defence committees. Yet a military imperative is often a factor and it is a formidable force. In the eyes of politicians and military leaders, there will always be a new enemy, a new reason to develop weapons, medicines, transportation systems and communication networks. From these have come, and will continue to come, new modes of war, and with them new bloody outcomes. But happily, as I hope this account will make clear, along this troubled road humanity has gained and will continue to gain things that are precious. For, as history makes clear, good as well as evil may flow from the darkest recesses of the human soul.

PART 1

FROM

THE GODS

TO THE

LASER SCALPEL

The Blood of Men

The Ionians of 1000 BC were the first to give a name to medical practitioners: they coined the word 'physician', which meant 'extractor of arrows'. But, although they had a name for those who tended the sick and injured, military leaders of the ancient world had little interest in what a doctor could do. For most commanders, a soldier wounded on the battlefield was a liability and his fate was usually left to the whim of the gods – to die or to recover; only high-ranking officers and noblemen were treated differently.

For the ancients, the services of the physician were often confused with those of the priest, which is why the cult of the witch doctor became a strong influence in the rituals and political framework of many early civilisations: surgery was linked with ritual and religious rite. The art of trepanning, a technique involving the drilling of a hole through the top of the skull (which is supposed to allow more air to the brain) had its origins in an occult ritual in ancient Egypt and it was a procedure performed by the priest-surgeon. And while such practices remained an aspect of the mystical canon for millennia, some early medics also began to accumulate useful, practical knowledge, both on and off the battlefield.

The quasi-mythical figure Aesculapius, dubbed the 'father of medicine', is thought to have been based upon a real person, a naval surgeon who lived during the twelfth century BC. He is said to have advocated high standards of hygiene and left descriptions of how best to extract arrows.

Many doctors believed they could be of active help to the military and were visionary enough to understand that experience on the battlefield provided invaluable training. Hippocrates himself wrote: 'He who wishes to be a surgeon should go to war.' But only a very few enlightened military leaders shared such enthusiasm or saw value in the presence of a token medic during their campaigns. Alexander the Great allowed medics to travel with his conquering armies but had little faith in them. He refused any medical help himself when he contracted a fever during the siege of Babylon in 324 BC and died soon after.

Records have survived of doctors who accompanied the Roman legions and left their mark upon conquered lands. One seventh-century Saxon treatment for sword wounds (which derived from an earlier, Latin script) suggested the surgeon 'Take the heads of iris and dry them very much and take thereof a pennyweight and a half, and the pear tree and rowan bark and cumin and a fourth part of laurel berries and of the other herbs half a pennyweight of each, and six peppercorns and grind all to dust and put two egg shells full of wine, then give it to the man to drink till he be well.'[1]

Such was the standard of most medicine throughout ancient times and through the Dark Ages in Europe, but there were a few inventive souls who did find ways to improve surgical technique and to alleviate the suffering of the soldier. A particularly visionary surgeon was Henri de Mondeville, author of a book entitled *Cyrurgia*, which he began in 1306 but never completed. This book, in which he opposed almost all traditional medicine, flew in the face of the second-century Roman medic Galen, who was then considered the all-knowing seer on the subject. Mondeville's approach was very modern and his advice concerning how best to treat wounds was often far ahead of its time. Describing his method for suturing the large intestine, he wrote: 'Sew as furriers sew a skin, before replacing it in the abdominal cavity.'[2]

Sadly, none of his techniques were adopted by his contemporaries and much of his work remained lost until it was rediscovered in the nineteenth century. However, Henri de Mondeville was probably the first doctor to apply to civilian life the knowledge he had acquired on the battlefield. Returning from campaigns in Aragon and Flanders, he set up a college of surgeons in Paris which he opened to the public on the first Monday of each month to treat '. . . common wounds and sores'.

By the time of the Renaissance many professional doctors were placing great value upon battlefield experience and using their learning for the benefit of civilians. The English doctor William Clowes made a name for himself as a fearless and effective military surgeon who saved the lives of hundreds of soldiers. In 1575 he was appointed to the position of chief surgeon at St Bartholomew's Hospital, London, and went on to write a treatise on wounds that became a standard for more than a century.

With the advent of printing, experienced surgeons who had witnessed more than their fair share of battlefield gore wrote accounts of what they had learned that reached far larger audiences than ever before. Hieronymus Brunschwig's *Buch der Wund Artzney* (*Book of Wound Dressing*), published in 1497, and Hans von Gersdorff's *Feldbuch der Wundartzney* (*Fieldbook of Wound Dressing*), published twenty years later in 1517, were read widely across Europe. Each illustrated how the types of wounds witnessed on the battlefield had worsened enormously thanks to the advent of gunpowder and the introduction of cannon balls and lead shot, but they also enabled those surgeons who had not experienced war first hand to apply new ideas to injured and sick civilians.

Perhaps the most far sighted of all military surgeons was the French doctor Ambroise Paré, who took part in his first campaign soon after completing his apprenticeship in 1536. Already hardened by his experiences as a medical trainee in Paris, Paré was nevertheless shocked by the conditions he discovered on the battlefield. In particular, Paré abhorred the practice of cauterising open wounds by applying boiling oil or a red-hot poker, believing that the technique caused more harm than good, not least because of the terror it engendered. One of Paré's sympathetic contemporaries, another young surgeon named Thomas Gale, later wrote: '. . . the heated irons so feared the people with the horror of cauterisation that many of them rather would die with the member on than abide the terrible fire by means whereof many people perished'.[3]

But as a junior doctor in the field Paré could do little but follow the orders of his superiors who, almost to a man, subscribed to the use of oil and hot metal. However, one night he found himself alone supervising a group of men who had just been brought in from the battlefield. When he went to the stores cupboard he found his supply of oil had been used up. After a brief moment of deliberation Paré decided to try something radical and dressed the men's wounds with a concoction he had been experimenting with – a blend of egg yoke, rose oil and turpentine.

'That night,' Paré later recounted in his memoirs,

> I could not sleep at my ease, fearing that I should find the wounded on whom I had failed to put the oil dead or poisoned, which made me rise very early to visit them where, beyond my hope, I found those upon whom I had put the digestive medicine feeling little pain, and their wounds without inflammation or swelling having rested well throughout the night; the others to whom I had applied the said boiling oil, I found feverish, with great pain and swelling about their wounds. Then I resolved in my heart never to burn thus cruelly poor men wounded with gunshots . . . I dressed and God healed.[4]

Buoyed up by this success, Paré went on to find a solution to another problem regularly faced by the military surgeon. Men with severe injuries to their limbs were frequently abandoned on the battlefield, but if they were rescued they were left for at least two days to see if gangrene killed them. If they survived this trial, they would then have to endure amputation without anaesthetic. Not surprisingly, most of the patients died from shock on the operating table, but many others died from loss of blood. Paré believed that the standard technique of cauterising severed blood vessels with a hot iron was not the best way to deal with the patient's limb stump after amputation. During a campaign in northern Italy he had devised an alternative. Using a crescent-shaped needle he pricked the blood vessels, led the needle through with a pair of pincers and knotted the ends of the thread tightly. In this way Paré invented the technique of suturing; this remained almost unchanged until the invention of the laser scalpel which can seal blood vessels and wounds on a microscopic scale using a distinctly more modern form of cauterisation.

Although Paré found that this technique sometimes required up to forty-eight individual ligatures or stitches, the value of his method was confirmed by the survival rate of those he operated on – three times as many amputees in his care survived surgery compared with those who underwent the traditional operation. But again, in spite of the fact that Paré went on to become one of the most acclaimed doctors of his age and was surgeon to four successive kings of France, it was not until the nineteenth century, three hundred years later, that his radical ideas were universally adopted. Indeed one of Paré's rivals from the esteemed Paris Faculty of Medicine went so far as to say of his innovative methods: 'An ignorant and misguided person has recently dared, from lack of knowledge, to reject burning the arteries with red-hot irons after the severance of the limbs and, flouting common sense, to substitute a new treatment, the ligature of the arteries without being aware that a ligature is far more dangerous than cauterisation with red-hot irons . . . In truth, anyone who endures this butchery has every good cause to thank God that he is still alive after the operation.'[5]

Such conflicting views between medics were not uncommon during the seventeenth and eighteenth centuries, and there was a vast difference in the level of sophistication between the ideas of visionary innovators and many establishment 'experts'. A comparison of two very different medical reports illustrates this difference very clearly. In 1689, an anonymous English doctor, describing his 'New Treatment for a Wound', wrote: 'Boil two pounds of oil of lilies, two new-whelped puppies till the flesh fall from their bones; add some earthworms in wine. Then strain and to the strained liquor add turpentine and an ounce of spirit of wine.'[6]

The very same year, one of Paré supporters, the French surgeon Morel, invented the first tourniquet, a device that would save countless lives both on and off the battlefield. Eloquently describing its use in stemming blood flow during an amputation, he reports: 'Place a rod of steel some four inches in length against the main artery. Secure with a band so tightly that the beat of blood below the ligature ceases. Now perform your amputation.'

But if the inventive techniques of men such as Paré often took many years to be accepted within the medical fraternity, the bravura of war-hardened surgeons certainly spilled over into civilian operating theatres. In 1824, the famous surgeon Astley Cooper was acclaimed for his ability to carry out an amputation in under twenty minutes. Cooper gained his experience from the battlefield. But within a decade others had dramatically improved upon this feat. The dashing and brilliant James Syme, who, as Wellington's chief field surgeon conducted hundreds of amputations in France, could sever tissue, bone and muscle and have a leg off in ninety seconds. In the pre-anaesthetic age, Syme's services were understandably much in demand.

Other methods and techniques that combined speed and great skill were brought back from the battlefield as well. Robert Liston, a lion of a man with a sharp tongue and a lightning temper, operated with ferocious speed and could be seen with the knife gripped in his teeth as he buried his hands up to the wrists inside a chest cavity or removed a tumour from an intestine. Another was Henry Cline, who was so dedicated that he lectured on his wedding day. A contemporary – and a man many considered to be the most influential of this generation of surgeons – was George James Guthrie, who demonstrated how limbs could be saved by removing a joint rather than an arm or leg. He also devised a splint for soldiers with thigh injuries, which was, like Morel's tourniquet, a simple but dramatically effective innovation.

To the modern mind, the very idea of surgery without anaesthetic is almost too horrible to contemplate, but before the mid-nineteenth century there was simply no alternative. The early nineteenth-century novelist Fanny Burney left a graphic and courageous record of her traumatic mastectomy, endured without anaesthetic: 'Mr Dubois placed me upon the mattress and spread a cambric handkerchief upon my face', she wrote. 'It was transparent however, so I saw through it that the bedstead was instantly surrounded by seven men and my nurse. I refused to be held, but when, bright through the cambric, I saw the glitter of polished steel – I closed my eyes – Yet, when the dreadful steel was plunged into the breast, cutting through veins, arteries, flesh and nerves, I needed no injunctions not to restrain my cries.'[7]

A sixteenth-century surgeon has left us with a similarly harrowing description: 'I was about to cut off the thigh of a man of forty years of age, and ready to use the saw, and Cauteries. For the sick man no sooner began to roare out, but all ranne away, except only my eldest Sonne, who was then but little, and to whom I had committed the holding of his thigh, for forme only; and but that my wife, then great with child, came running out of the next chamber, and clapt hold of the Patient's Thorax, both he and myselfe had been in extreme.'

The first form of anaesthetic was formulated not by battlefield surgeons but rather within the rarefied academic atmosphere of the Royal Society, where Humphry Davy first conducted experiments using nitrous oxide. In 1800 he described his findings in a book, *Researches, Chemical and Philosophical, Chiefly Concerning Nitrous Oxide and its Respiration*, but these experiments had almost no impact upon the medical world and nitrous oxide was later only rarely used in experiments with human subjects.

The power of chloroform (which might be considered the earliest anaesthetic accepted within the medical profession) was first noticed almost simultaneously in the United States and Europe in

1831. The American Samuel Guthrie, the German chemist Justus von Liebig in Germany and the French medic Eugène Soubeiran each experimented with chloroform and wrote accounts of its seemingly miraculous properties. When he first witnessed the use of anaesthetic, the physician Johann Dieffenbach declared: 'That beautiful dream has become a reality: operations can now be performed painlessly.'[8]

But chloroform had a darker side. Within a few years, its use had fallen out of favour with those few doctors who had tried it. The chemical certainly worked in anaesthetising the patient but it also caused many of them to suffer heart attacks.

By the mid-1840s ether had replaced chloroform as the only usable anaesthetic, and a few years later a more effective agent, ethyl chloride, was adopted. Yet even as late as the 1860s these anaesthetics were not known widely. Queen Victoria had popularised the use of chloroform when she allowed its use during the birth of her son Leopold, but when chloroform fell from favour doctors were slow to accept alternatives. This situation was not helped by a series of botched attempts to use the new technique, combined with bad publicity. One Massachusetts dentist, Horace Wells, did little for the image of anaesthetics when he miscalculated a dosage and caused his patient agonies during a demonstration before a dentistry class. Wells was discredited, grew depressed and ill and became addicted to chloroform. In a delirium, he threw sulphuric acid at two prostitutes; he was arrested and subsequently committed suicide in jail. The story made the front pages of newspapers from coast to coast, delaying by years the acceptance of anaesthetics.

A complete reversal in attitudes towards anaesthetics came with the American Civil War. At the Battle of Gettysburg alone the two sides suffered 54,807 casualties and, during General Grant's offensive at Coldharbour in 1864, ten thousand men were wounded in a single hour. These soldiers either died where they lay or were operated on in makeshift operating rooms. The surgeons, covered in blood, wiped their knives on their filthy aprons as they moved from soldier to soldier, removed an arm here, sewed up an intestine there. Thankfully, chloroform and ether were available to some patients, but supplies were soon exhausted and many of the wounded died from shock.

After this war, all armies went into battle supported by a medical division well stocked with anaesthetics. Soldiers who survived because of ether and ethyl chloride and could remember the horrors of the pre-anaesthetic age came home enthusing about the wonder chemical that had helped to suppress their agonies. As a consequence, by the end of the 1860s, all operations in both the United States and Europe were conducted under anaesthetic.

The American Civil War proved to be a watershed for the military medics. The popularising of anaesthetics was perhaps the most important benefit to come from that war, but other innovations also saved many lives. Medical logistics were greatly improved by the first use of efficient carts and stretchers to ferry the wounded from the battlefield, field hospitals were better run, supplied and staffed effectively, and war surgeons devised a specially designed bullet probe, or forceps, that could be used to remove tiny fragments from deep wounds. This device soon found ready application in civilian hospitals.

The First World War is considered by some to have been a war that bridged the ancient military world and the modern. Like all wars, it was filthy and squalid and it degenerated human beings to the level of animals driven to slaughter, but this time on a previously unimagined scale. The battlefields of the Somme, Ypres, Flanders and Passchendaele displayed all the horrors of past conflicts but magnified a hundredfold by the sheer volume of casualties. It was also a war in which new, more deadly weapons were used, weapons that offered the surgeon and the medics fresh and daunting challenges.

During the Second World War these problems were exacerbated by the advent of more powerful rifles and because land mines became more commonplace and increasingly sophisticated. One particularly unpleasant example was the anti-personnel mine deployed by the Germans in North Africa. Roughly the size of a jam jar, it was activated by being trodden on, whereupon it sprang five or six feet into the air, exploded and scattered three hundred metal balls.

Such a trend continues still. Evolving through the Vietnam War and the two Gulf Wars guns, cannon, anti-tank devices and mines have become increasingly deadly, so that soldiers injured on the battlefield arrive on the operating table suffering from many different life-threatening wounds. Data collected by the US Army during Operation Desert Storm shows that the *average* number of wounds suffered by soldiers admitted to field hospitals during the conflict was one hundred.

Terrible injuries have led battlefield surgeons to take drastic measures that have produced unexpected benefits. During the Second World War, a researcher at Eastman Kodak named Harry Coover developed a transparent glue made from a group of chemicals called cyanocrylates to be used to repair damaged gun sights. In Vietnam, surgeons began to use this glue in the form of a spray to seal up potentially fatal wounds, giving them time to get the injured to a field hospital. After this success in Vietnam, the material continued to be used in medicine but also found another, more widespread outlet as Superglue.

More recently, the US Defense Department has been developing a unique bandage called the Chitosin bandage, designed by the Medical Research and Materiel Command, which is capable of stopping severe arterial bleeding within two to four minutes of application. The adhesive nature and enhanced clotting capability of this bandage provides wound pressure and bleeding control to external haemorrhages. According to the US Army's chief scientist, Thomas Killion, in a government report emphasising the enormous value of this innovation, the bandage may be '. . . utilized successfully on a variety of injuries ranging from gunshot wounds to landmine injuries. Bottom-line . . . it saves lives.'[9]

A ghastly escalation in the severity of war wounds is inevitable. It demonstrates the huge energy and ingenuity expended upon the development of increasingly vicious weapons and fills one with horror at the prospect of what might lie in the future. But, if we can see a silver lining in this dark cloud it is the fact that, as the

injuries become more complex, the surgeon learns more. The first
effect of this knowledge is to nullify some of the power of the
weapons and the injuries they cause, but also to increase the flow
of information into civilian medicine.* Most striking is the fact
that many of the injuries caused by modern weapons resemble
facial and limb injuries suffered by road-accident victims.

The advances that have come from treating battlefield
wounds – new forms of lightweight splints, more effective
painkillers, better bandages that allow more efficient circulation of
air, better designed monitoring equipment, and, crucially, rapid
improvements in surgical technique – have all greatly improved
civilian emergency services and the survival rate of accident victims. But by far the most important advance was a technique that,
like the use of anaesthetics, had been theorised about and practised *ad hoc* for many years but only came to the fore through its
use in field hospitals.

Samuel Pepys witnessed some of the earliest attempts to trans-
fuse blood. He was in attendance when Christopher Wren tried
injecting a variety of liquids into the bloodstream of a dog, and
later he recounted the ham-fisted efforts of a French physician,
Jean-Baptiste Denys, to pass blood from one dog to another. On
another occasion, in 1667, Pepys, a Fellow of the Royal Society,
saw Richard Lower conduct a blood transfusion from a sheep to a
divinity student in his care, one Arthur Coga, who was described
as being 'somewhat crack-brained'. The change of blood, the noble
Royal Society Fellows believed, might offer the lad an end to his
suffering. In a way it did: within minutes he died before the enrap-
tured gaze of his audience.

On the battlefield there had been many attempts to pass blood

---

*An illustration of how medicine has usually managed to keep one step ahead of the
weapons developers comes from a survey of the survival rates for soldiers wounded
during the Second World War compared with the numbers during the First World War.
Death from field hospital amputations dropped from 70 to 20 per cent, mortality from
thoracic injuries plummeted from 54 to 5.7 per cent, and between 1939 and 1945 only
37 per cent of eye wounds led to blindness, whereas, during the earlier conflict, 67 per
cent of soldiers who suffered eye injuries lost their sight.

from a healthy soldier to one bleeding to death on the operating table. The first attempt was made by a British doctor named Blundell in 1818; and in 1870 at the Battle of St Privat, during the Franco-Prussian War, casualties were so numerous that doctors were driven to attempt transfusions in the mud in which it is said five thousand Germans were wounded in fifteen minutes. Not a single soldier survived these desperate efforts.

The reasons for the failure of these experiments would seem obvious to modern eyes but nothing was then known of different blood types. Indeed, it was to be another two and a half centuries after Pepys's experiences before researchers began to understand the many roles of blood and the complexity of its composition. The pioneer in this field was a Viennese physician named Karl Landsteiner, who spent years studying the data from hundreds of failed transfusion tests until he gradually began to realise that not all blood was the same. Blood, he suggested, carried different types of agents called antigens (chemicals that produce defensive agents or antibodies that help the body to fight infection). Landsteiner discovered that if blood containing different antigens was mixed together, red blood cells would clump together and kill the patient. Only blood containing identical antigens could be mixed or transfused from one individual to another. As a result of his researches, Landsteiner surmised that there are four main types of blood, which he called types A, B, O and AB. Based upon this supposition, he could explain why it was that, on rare occasions, transfusions had proven successful: by chance, the experimenter had used subjects with the same blood group.*

It was only as war in Europe seemed inevitable and both Britain and the United States began to prepare for the conflict to come, that some forward-thinking military medics turned their thoughts to the nature of blood and the possibilities of blood transfusions.

---

*This is actually a rather simplified account because group O comes in two forms, O positive and O negative which are mutually incompatible. This incompatibility was only discovered in 1939. Also, blood group 0 can be transfused to people with the other groups, which may account for many of the 'rare' successful transfusions before the science of blood was understood.

By this time, Landsteiner was working at the Rockefeller Institute for Medical Research in New York, and during the final few years of his life (he died in 1943 at the age of seventy-five) he spearheaded the drive to build up blood banks and to devise methods of transporting blood from civilian donors to the front line.

It is impossible to estimate the number of lives saved by the advent of blood transfusion both on the battlefield and within the civilian sector. The Red Cross did more than any organisation to provide blood plasma for the war effort, transporting an estimated thirteen million pints to the battle front; and, after the war, they conscientiously kept the public informed about the need for blood banks.

Today, special units track the reports of field doctors and surgeons and innovation is encouraged. Often, surgeons returning from war zones write papers and books describing new techniques and methods they have stumbled upon while operating on the wounded. This knowledge very quickly filters back into everyday medicine.

For many centuries the findings of innovative surgeons on the battlefield were ignored but this is now far less likely to happen. For the military surgeon great kudos may be gained from making an important advance, and anything that works in war, it is now believed, may be refined and made valuable in civilian life.

The Fighting Man's Panacea

If the slash of a sword or the thud of an arrow brought sudden pain and the likelihood of imminent death, the ravages of disease, ever fertile in the slime and gore of battle, often heralded a slower but equally inevitable end.

As late as the 1920s, more soldiers died from disease, poor diet and exposure than from wounds. The Union Army in the American Civil War lost 186,216 men to disease, twice the number killed in action, and of these 45,000 men died of dysentery alone. During the First World War, the big killer was typhus, a disease that in Serbia killed 150,000 soldiers during the first six months of the conflict and had taken three million Russian lives by the end of the war. This catastrophe led Lenin to comment: 'If Socialism does not defeat the louse, the louse will defeat Socialism.'

Until medieval times wars were often fought in a detached environment far from civilian life. Peasants and noblemen alike saw their kinsfolk leave to fight for the king or queen in distant lands that took months to reach. Often, they waited in vain for their return. But by the early Middle Ages, wounded soldiers were sent to monasteries to recover and nuns and monks became the first nurses. These men and women devised their own methods for treating wounds and sickness. And, when the battles had passed on and the soldiers were either buried in the grounds of the monastery or discharged to fight another day, the techniques learned from their treatment were applied to help the community.

An odd but influential aspect of this relationship was that most people (including men and women of the cloth) considered a soldier incapacitated by disease to have suffered a 'less honourable' fate than those injured in battle. This strange belief stemmed from the idea that all forms of disease were visited upon the unlucky soul as a punishment from God, a retribution for some ill-defined sin. However, a wound received from an enemy sword or arrow was deemed 'righteous' and the victim treated with honour. As an unfortunate consequence, knowledge of how best to treat wounds far outstripped ways of dealing with infection, even though diseases such as plague and those caused by poor sanitation and inadequate diet took the lives of many more soldiers than those struck down in battle.

The trials of the soldier often only began with a wound. Anyone unlucky enough to be struck by an arrow or hit by shrapnel was at the mercy of the surgeon's skill in removing often deeply buried shards. Richard Cœur de Lion died at the siege of Chalus in 1199 because no less a medic than the royal surgeon was unable to remove an arrow without rupturing an artery. But even if a soldier survived surgery they were still prey to the capriciousness of Nature itself, the random flow of bacteria and virus. Before the advent of antiseptics and antibiotics even a relatively minor laceration could lead to infection and death.

Few diseases were more ruinous to an army than syphilis. This bacterial infection was the AIDS of its day and it stalked civilisation for almost half a millennium, killing millions. Columbus brought the disease back with him from the New World to Europe and for a time the looming disaster went unnoticed; in the taverns and brothels of Isabella and Ferdinand's Spain, however, its invisible agents were soon at work. Within two years, syphilis had become as feared as the plague and every bit as deadly.

In 1494, a year after Columbus's return, Charles VIII of France attacked the Italian peninsula. His was a pan-European army, a pack of mercenaries made up of 18,000 French horsemen and 20,000 infantry from half a dozen countries, including 3000 pike-wielding Swiss foot soldiers. They crossed the Alps and swept into Italy almost unhindered. Within two months, Milan, Florence and Rome had fallen. The invaders were in festive mood and with good reason. They celebrated long and hard, drinking and whoring.

And after the first round of celebrations was over, exhilarated and confident, Charles's massive army moved on to Naples, where, with one exception, resistance melted. That exception came in the shape of a group of men who had retreated to a citadel just outside the city. They were a ragtag band of Neapolitan fighters who had little hope until they were joined by a contingent of Spanish soldiers, fresh from the taverns of Madrid and Seville. They brought with them much-needed weapons and supplies; they also brought syphilis.

Laying siege to the citadel, the French troops and their mercenaries simply sat and waited and caroused with every expectation that the Italians and Spanish would either surrender or starve to death. And the situation facing the citadel's defenders became bleak indeed. Food and fresh water ran short and, as the weather grew warmer, the cramped, unsanitary conditions inevitably led to disease. But then, as things reached a critical point, the besieged troops somehow found a way out of the citadel unnoticed by the French, and, as they planned their escape, they made a decision that would ensure the death of most of their enemies but would, in the fullness of time, also turn upon their own countrymen.

The anatomist Gabriello Fallopio, whose father was in the fortress, left an account of the final days there:

Since they were a small band, vastly outnumbered by the French, they stole out of the fortress, leaving behind an adequate garrison, and poisoned the wells. Not satisfied with this, they bribed the Italian millers who delivered the corn to the enemy to mix plaster in the meal, and finally, under the pretext that food was short, they expelled from the fortress, the whores and women, especially the attractive ones whom they knew were infected with disease. The French seized with compassion for the women and attracted by their beauty, gave them asylum.[1]

The dreadful ruse worked: the French were suitably distracted and were led to believe that all the civilians left in the citadel had died and that the Italian and Spanish soldiers had escaped to regroup. But, in fact, the military and political tide was turning against the invaders. Angered by Charles VIII's rampage, the Spanish and the Germans revoked their neutrality and threatened to attack France. At the same time, the Venetians, determined to oust the enemy from the peninsula, formed the League of Venice which began to supply Milan with arms.

Intimidated by the actions of his powerful former allies, Charles decided to retreat, but the League of Venice in the form of an army forty thousand strong was waiting for him and his army at Parma. The French, grown flabby and overconfident in the Neapolitan sun and with many foot soldiers and horsemen already falling sick, had little chance. The army was ripped apart, their discipline broken. A ramshackle band of survivors made their sorry way home across Europe to France, Switzerland and Austria.

Along the slow, torturous way, few gave them solace. Many soldiers fell sick and died by the wayside, others staggered on appearing as lepers, their faces and hands covered with syphilitic sores. In Alpine villages and along hot Swiss roads, once proud and triumphant French soldiers became blind and insane before finding the comfort of death.

The symptoms of some of the others, though, developed more slowly. Hundreds made it back to their homelands with only flesh wounds and almost imperceptible sores inside their mouths or across their nostrils, signs that might have been easily mistaken for cold sores or ulcers. Relieved to be home, these survivors passed on their deadly infection before the later stages of the disease gripped them too.

For centuries, doctors remained utterly powerless to stop the ravages of syphilis. It was generally believed that symptoms could be alleviated by careful administering of arsenic, while other quacks tried leeches, bleeding and secret remedies of their own. No one at this time had the faintest notion how a disease was transmitted and the very idea of 'infection' was an alien one. Often surgeons went straight from an autopsy to a childbirth without washing their hands and disinfectants were actively frowned upon by many.

By the 1830s, the idea that 'dirt' somehow aided the spread of disease was beginning to be accepted by many medics, but each doctor and every nurse had his or her own standards. An American doctor, Edward Jarvis, working in the Midwest during the 1830s, recalled in his memoir an incident that illustrates this confusion perfectly. While helping out at a colleague's practice, Jarvis was supervising the bandaging of a wounded leg. He turned to a medical student and asked for a plaster. 'I do not know if we have exactly the one you need in the drawer,' the young man replied. He then opened a cupboard and Jarvis saw to his horror that inside lay an array of plasters and bandages of all types that had been removed from sores, ulcers and cuts, many caked in blood and pus. They had simply been piled into the cupboard for future use.

By the time of the Crimean War in the mid-1850s it had become evident that infections on the battlefield and in field hospitals were the cause of many otherwise inexplicable deaths. In particular, the 'syphilis problem' was threatening to lose the war for Britain. Doctors were aware that somehow syphilis was most prevalent among prostitutes and the disease was being passed on to the fighting force by this route, and so the government took firm action, banning the transport of prostitutes from England to the front and stopping foreign prostitutes or those known to have the disease from entering England. Similar measures continued after the war. In an attempt to curb syphilis, prostitutes in London, Manchester and other major cities were subjected to regular medical examinations and were imprisoned if they were found to be diseased.*

---

*Yet, even these Draconian measures did little to reduce the power of venereal disease. The scourge of syphilis was only ended when antibiotics became generally available, which, as we shall see, came directly from another war development during the 1940s.

A similar awareness of the cause of disease came from lessons learned during the American Civil War. Of the six hundred thousand Civil War dead, two-thirds had succumbed to disease, a statistic so shocking that the American government was prompted to create the Civil Sanitary Commission and to appoint at its head the New York physician Elisha Harris. This revolutionary organisation immediately brought about reform of the hospital system and improved the medical care facilities and diet of the poor.

By the second half of the nineteenth century, doctors and scientists had slowly begun to find clear links between disease, hygiene and diet. Through the new disciplines of biochemistry and other life sciences, researchers were revealing the true source of infection to be bacteria and viruses. At the same time, society's attitude towards treatment and disease was changing. The idea that a disease was the result of God's vengeance or because of the insidious influence of evil spirits was rapidly losing favour and the potential of science and discovery was embraced and encouraged.

The Scottish Quaker Joseph Lister embodied this new spirit perfectly. His father was a successful academic who had been involved in the development of the microscope, and Joseph was sent to University College to study medicine. Soon after his graduation in 1854 he was appointed as an assistant to the famous Edinburgh surgeon James Syme, whose daughter he later married. As a surgeon, Lister learned quickly the devastation caused by infection.

Sepsis was one of the biggest killers in the hospitals of that time. In some institutions childbirth mortality rates from sepsis infection rose as high as 30 per cent, and the disease killed almost 80 per cent of amputees within days of surgery. These horrifying statistics prompted one surgeon of the day to admit candidly: 'Those entering hospital for surgery are exposed to more chance of death than the English soldier on the field of Waterloo.'

The medical world had also been stunned by the number of soldiers who had died from disease during the Crimean War, and the newspaper editors of the day expressed outrage that so many young men were dying in war not from the effects of enemy guns, but from illness.

Academically brilliant, Lister was also well read and open to new ideas. As a student he had been fascinated with medical history and so he was quite aware of past research into the mechanism for infection. He referred to Pasteur's work to elucidate the role of bacteria in the processes of rotting and fermenting as 'the Frenchman's beautiful research'. Following Pasteur's lead, by the late 1850s Lister had become convinced that bacteria caused most of the infections acquired during surgery and postoperatively, and he concentrated his efforts upon developing a practical antiseptic and aseptic technique.

Soaking specially prepared bandages in carbonic acid, he used these to bind the wound and a metal foil, or a 'Macintosh' (a form of rubber), was placed over the bandage and sealed tightly. This arrangement had the effect of producing an impermeable membrane to keep out bacteria. Taking the idea further, he adopted the controversial step of having his operating theatre sprayed with carbolic throughout an operation and went to great lengths to make sure the instruments, the bandages and the hands of the surgeons were scrupulously clean.

These measures produced immediate and striking results. Before introducing his antiseptic and aseptic methods in 1866, Lister's record showed that of thirty-five amputations conducted between 1864 and 1866, nineteen patients lived and sixteen died, a mortality rate of 45.7 per cent. After his techniques were introduced, the next forty operations conducted over a period of three years led to thirty-four survivors and six deaths from infection, a mortality rate of just 15 per cent.

And yet most of the medical community either ignored Lister's results or attempted to discredit them. Unable to accept the heresy that doctors themselves could be responsible for infecting patients and that invisibly small entities could provide the means, one eminent professor of surgery, John Hughes Bennett, declared: 'Where are these little beasts? Show them to us, and we shall believe in them. Has anyone seen them yet?'[2]

And so, just as the civilian medical profession had been slow to grasp the importance of anaesthetics or the revolutionary ideas of men like de Mondeville, Paré and Guthrie, it took military imperatives to hammer home the value of Lister's radical scientific findings.

Lister's methods came too late for most soldiers who fought in the Franco-Prussian War of 1870. Of the 13,200 men who had limbs amputated, ten thousand died of gangrene and other infections, a staggering mortality rate of 76 per cent. Many of the surgeons who witnessed the horrors of this war were then understandably open to Lister's revolutionary ideas that hovered on the fringes of medicine. One doctor fresh from the front and desperate to reduce his postoperative mortality rate was a young German, Johann Ritter von Nussbaum, who visited Lister in Scotland and returned home extolling the virtues of the great surgeon's ideas. 'Behold now my wards,' he enthused, '. . . which so recently were ravaged with death. I can only say that I and my assistants and nurses are overwhelmed with joy and gladly submit to all the trouble this treatment involves.'[3]

Others were equally impressed and were willing to conduct tests in their own hospitals. Across Europe, adventurous doctors who adopted 'Listerism' soon saw mortality rates on their own wards plummet in spectacular fashion. Then, a few years after the Franco-Prussian War, in 1878, Pasteur demonstrated before the French Academy of Medicine a clear argument for his germ theory of infection before publishing a landmark paper providing the theoretical support for Lister's practical findings. Within a decade, thanks to a combination of practical and theoretical ingenuity propelled by the impetus of mass slaughter on the battlefield, surgery was totally transformed.

Even so, there was still much room for improvement outside the operating theatre. On average, 15 per cent of surgery patients died within three days of an operation, and each year large numbers of people succumbed to bacterial and viral infection; a sore throat or an infected cut could kill. Clearly, a more effective attack against infecting agents than chemical sprays or bandages soaked with carbolic acid had to be found.

The lion's share of the credit for the discovery of the first anti-biotic, penicillin, has always been reserved for the British medic Alexander Fleming, but in some ways this is misleading. Fleming, a rather glamorous upper-class doctor who did a lot of his best thinking on the polo field or while out hunting, was instrumental in the discovery, but many others of equal ability played crucial roles in the breakthrough.

In 1900, at the age of nineteen, Fleming volunteered for the British Army to fight the Boers in South Africa, where he witnessed some of the most barbaric fighting of his or any age. Upon his return home, he entered medical school using money from an inheritance to finance his course and, a few years later, found himself fighting again, in the First World War, this time as a member of the Army Medical Corp. It was during this conflict that Fleming first realised the true inadequacies of some of the antiseptic techniques then in practice. He noticed that the chemical agents used to fight infection often lowered the body's own defences and some failed to destroy the bacteria that cause infection. Furthermore, antiseptic methods that had changed little since Lister's day did absolutely nothing for soldiers with influenza, typhoid and a host of other debilitating diseases.

Back home, Fleming's thoughts were focused even more sharply by the terrible influenza pandemic of 1918–19 (a plague that, in the space of two years, claimed an estimated twenty-two million lives), and in 1921 he made his first major discovery. A Petri dish in his lab containing a sample of nasal mucus had been contaminated with an enzyme called lysozyme, found commonly in tears. On the lookout for substances that destroyed bacteria, Fleming quickly noticed that where the lysozyme and mucus met, the mucus sample had been dissolved. He concluded that lysozyme might be one of the substances the body employs to defend itself against infection.

This was an important step in the right direction, but lysozyme itself did not attack harmful bacteria; it simply acted in unison with other substances as part of the body's natural immune system. For the next six years, Fleming worked with various strains of bacteria, in particular the type known as *staphylococci*, which he knew was responsible for boils, abscesses, septicaemia and pneumonia. In August 1928, Fleming returned from a holiday to find that one of his Petri dishes containing the *staphylococci* bacteria had a mould growing over it, and where the mould had proliferated, the bacteria were dead.

Fleming identified the mould as *Penicillin rubrum* (actually, he got this slightly wrong; it was a close relative, *Penicillin notatum*). This mould produces a chemical that is lethal to many bacteria including *streptococci*, *gonococci*, *meningococci* and *pneumococci*, but at the same time it leaves healthy cells unaffected and does not impede the defensive functions of white blood cells (*leucocytic* cells).

Fleming understood immediately the implications of what he had discovered and he wrote a paper describing his findings. Unfortunately, his researches ended with this paper because, when he took the next step – an attempt to refine a form of penicillin extracted from the mould – the material not only proved chemically unstable but it could only be produced in vanishingly small quantities. Disillusioned, Fleming turned away from his line of research and never returned to it.

And that might have been the end of the penicillin story. But a full ten years later, an Oxford team headed by the Australian Howard Florey and the eminent German émigré biochemist Ernst Chain (along with Norman Heatley) stumbled upon Fleming's long-forgotten paper and began to extract their own samples of penicillin from the same kind of mould Fleming had used.

Within weeks they too encountered the same stumbling block Fleming had met – they could produce only one part in two million of the original mould. But unlike Fleming, they persisted and considered other forms of the penicillin mould from which they hoped to achieve better yields. Using a close relative of *P. notatum*, Florey and his team were able to produce just enough refined material to conduct a few tests on laboratory mice that they had infected with *streptococci*.

These first tests, conducted on the evening of 25 May 1940,

were a complete success. Four infected mice were inoculated with penicillin and four others (also with the *streptococci* infection) were left untreated. The next morning, all four inoculated mice were still alive; the others were dead.

Excited by this success, the team started producing as much penicillin as they could and simultaneously began searching for a suitable human test candidate. A few months later they had produced what they believed to be sufficient quantities of antibiotic from home-made equipment – vats, steel baths and milk churns filled with mould and solution – and before long they were given their first chance to test its powers on a human patient.

A London policeman named Albert Alexander had scratched his finger while pruning his rose bushes. The cut had become so badly infected that within three days he had a temperature of 105°F and was close to death. He was suffering from staphylococcal septicaemia. Albert Alexander was injected with penicillin and began to show immediate improvement. The course was continued for three days. But then, just as the policeman regained consciousness and appeared to be on the road to recovery, supplies of the drug ran out. Within twenty-four hours, the bacteria had again taken hold, the patient had relapsed into a coma and he died shortly afterwards.

It was a sad and disappointing outcome but it laid the foundation for a huge medical success. Although Alexander had died, the drug had clearly been effective. Indeed, the incident gave rise to an example of wartime gallows humour that quickly become a catch phrase among doctors: '. . . The treatment was a success, but the patient died.'

It was now 1941 and Europe was ravaged by war on two fronts. Florey understood that the only way to make penicillin an effective drug was to find a way to mass-produce it. He approached the British government and pharmaceutical companies, but, with wartime resources already stretched to breaking point, help was not forthcoming.

But Florey, supremely confident he was on to something very big, was determined to push on. The Oxford workers convinced

the American government of the importance of their research and
they were soon ensconced in a new complex of laboratories at the
US Department of Agriculture near Washington, DC, and the
National Regional Research laboratory in Peoria, Illinois. Dozens
of biochemists and hundreds of chemical engineers were recruited
to concentrate their entire efforts into finding a way to increase
the yield of the penicillin mould. The US government immediately designated the effort an Official War Project and gave it top
priority.

After going down a series of frustrating blind alleys and per-
forming many unsuccessful experiments, the team finally found a
mould called *Penicillin chrysogeum* that produced two hundred
times more extract than *P. notatum*. But this yield could still barely
meet the needs of the war effort and so the team continued to
try thousands of different forms of mould. By 1942, soon after
America entered the war, the researchers decided to take the
radical step of irradiating the mould, hoping to form a more
productive mutant fungus. The idea worked better than anyone
could have imagined and within weeks the Peoria laboratory was
producing thousands of times more pure penicillin than the original moulds had been capable of.

It was now a race against the clock. Through 1942, researchers
in the US and Britain (by now a few companies there had also
begun producing the irradiated mould) watched powerless as,
each day, thousands of soldiers and civilians died from infections
that could have been treated if adequate supplies of penicillin had
been available. By early 1944 (when production of penicillin had
been taken on by a dedicated plant built by Pfizer) large quantities
of the drug could be produced very quickly so that by the time of
the D-Day landings of June 1944 there were ample supplies of
penicillin for all the field hospitals and medical teams involved in
the invasion. After the war, penicillin became known as a 'wonder
drug', and, indeed, it has saved and continues to save millions of
lives each year.

During the past decade it has become evident that many forms
of bacteria are now resistant to antibiotics. To remedy this

problem, patients are being treated with increasingly powerful blends of antibiotics.

The importance of penicillin and antiseptics and an increased awareness of hygiene in wartime and in everyday life is almost immeasurable. These discoveries have changed medicine dramatically and improved all our lives enormously. Life holds many dangers and Nature still offers constant challenges for humanity, but at least we no longer need to rely upon the will of the gods to defend ourselves against bacteria.

However, this is not the end of the story; science and technological progress are rarely that simple. The story of how mankind has learned about micro-organisms and developed defences against their effects is a perfect example of the symbiosis between war and science. But, as illustrated by the atomic research carried out at Los Alamos during the Second World War, there is a constant feedback between science and conflict.

As mentioned in the Introduction, the building of the atomic bomb came from the work of theoreticians and experimentalists of the 1920s and 1930s, physicists such as Schrödinger, Einstein, Bohr and Curie. Their work was then exploited by military strategists and politicians for their particular ends. This exploitation resulted in the bombing of Hiroshima and Nagasaki, the nuclear arms race and the proliferation of atomic weapons. But the scientific knowledge that came from Los Alamos has also fed back into nuclear technologies far removed from military goals, and, as we shall see, has provided society with an array of technological advances, from the laser to the digital computer.

In the same way, research that led to an understanding of microbes has given us antiseptics and penicillin, but these discoveries have also been applied to the production of biological weapons. In learning how to prevent infection, scientists also learned how to cause the proliferation of disease, how to enhance the function of micro-organisms and how to refine this technology for wholly destructive purposes.

Biological warfare is no new thing; the Mongols of the thirteenth century catapulted rotting corpses into enemy forts and the early

British settlers in America murdered indigenous peoples by presenting them with blankets deliberately infected with smallpox. But it was only after the mechanism of infection was understood that it could be perverted to create the potentially devastating biological weapons of modern times.

Yet, even this evil has a beneficial side; for, as billions of dollars are poured into biological weapons research around the world, scientists learn more about the life forms they are dealing with and this knowledge is returned to non-military research projects in precisely the same way that knowledge from the Manhattan Project later fuelled postwar technology.

During the first Gulf War, all American and British troops were injected with a cocktail of anti-bio-weapons agents and anti-nerve drugs (including pyridostigmine bromide, the compound currently at the centre of the controversy over Gulf War Syndrome). Arguments over the side effects of these chemicals still rage and will continue to do so for a long time, but a little known benefit from this controversy is the enormous amount of research that has been conducted on both sides of the Atlantic into the effects of anti-bio-weapons agents and anti-nerve drugs. This research has led to extremely valuable insights that have provided medical scientists and biochemists with information about how the human body reacts to compounds such as pyridostigmine bromide, how the vastly complex mechanisms in the nucleus and cytoplasm of individual cells respond to alien agents and how delicate processes governing the function of the immune systems operate. Such knowledge takes our understanding of infection and the immune systems into previously unexplored areas, and it is hoped that one day these investigations will help scientists develop effective anti-influenza and anti-cold vaccines, and possibly even vaccines against and a cure for AIDS.

The threat to the survival of humanity posed by bacteria and viruses is matched only by our determination to hurt each other. It is perhaps ironic that the enemy within – our ceaseless aggression – has led to and will continue to lead to ways to keep at bay the enemy from without.

The Feminine Touch

It has been said that 'men start wars and women pick up the pieces'. In almost all cases this is certainly true, but in the ancient world no one was there to pick up the pieces.

Even as recently as the fourteenth century, the role of the nurse was a confused confection of priestess, servant and medic. This blurring of roles came from a conviction that medics possessed mystical powers beyond the scope of ordinary morals. One historian has said: 'It was believed that medical helpers possessed more than mortal knowledge, a knowledge ascribed to a pure unearthly being embodied in the ideal woman. The knight, who felt to his heart of hearts the charm of her beauty, was not slow in believing that she could fascinate the very elements of nature to aid him.'

Within a few centuries, this surreal image of the nurse or female medic had become rarefied to the point where the only females who could treat the wounded or sick were nuns, and care for the sick or wounded could only be found in monasteries. Beyond the sacred walls, any woman involved with serving the sick was perceived as being little better than a prostitute, an unfortunate misconception that hampered any progress in the field of nursing for centuries. Indeed, as late as the 1850s, when Dickens portrayed nurses as drunken hags such as Betsy Prig and Sarah Gamp, he was merely modelling his characters on an existing stereotype.

But in the nineteenth century three women, two American and one British, were to change entirely the public view and the role of nurses, and each did so at the forefront of battle. Horrified by what she saw during the American Civil War, Dorothea Dix, a respectable, middle-class woman (who at the start of the war had travelled to the front under her own steam), campaigned vigorously for the establishment of an officially sanctioned nursing corp. By 1861, she had convinced the government to fund her goal and she was appointed Superintendent to the United States Army Nurses. Something of a blue stocking and with deep religious convictions, Dorothea Dix had well-defined ideas concerning the image a nurse should have. Perhaps in an effort to reverse the outmoded prejudices facing her profession, she recruited '. . . only very plain women . . . no curls, no jewellery, and no hoop skirts'. But as the number of wounded in the battles of the Civil War escalated she was forced to relax her rules a little to allow for 'unadorned pretty nurses'.

One of her most devoted assistants was Clara Barton, who became perhaps the most famous nurse in American history. Known as the 'angel of the battlefield', Barton witnessed more than her fair share of torment during the Civil War and afterwards she campaigned vigorously for reform of both the civilian hospital system and the medical treatment of soldiers. But when she found she was being ignored by both the military and senior figures within the medical establishment, Barton took matters into her own hands and established the American Red Cross. In 1882, she almost single-handedly pushed through what became known as the 'American Amendment' to the mandate of the 1864 Geneva Convention, a statute that empowered the Red Cross to operate in civilian life as well as during times of war.

A few years before the American Civil War, Britain had fought an equally cruel and bloody conflict in Europe. The Crimean War of 1854–6 involved an alliance of the British, French and Turkish against the Russians over disputed territories close to the border between Europe and Asia. Because this war was fought during a period of great industrial expansion by nations that were more open to technological advance than at any earlier time, lessons learned from the conflict fuelled progress in many different areas.

We have seen how in Britain the effects of the Crimean War changed the attitudes of doctors towards the treatment and prevention of syphilis and how information about infection and disease encouraged Lister to develop his theories. But an equally important by-product of this war was the emancipation of women in medicine and the shift in the public's perception of professional nurses. No other individual effected this change more than the near-legendary Florence Nightingale.

Florence Nightingale was born into a wealthy family in 1820. She had been conceived in Italy while her parents were touring Europe and she was named after the Tuscan capital. Tutored by her father at home, her early life was typical of the cloistered and claustrophobic existence of upper-class women of the day. She was expected to marry well and do little else but act as gracious hostess and devoted mother. But, although she was pious and loyal to her family, Florence was not so easily moulded by convention. She later described how, in 1837, at the age of seventeen she had been walking in the garden of the family home in London when she had a vision of God and a heavenly voice told her that her true role in life would be to help the sick and the needy.

But the strict moral and social codes of the day made it extremely difficult for Florence to do anything practical to fulfil what she now believed to be her calling, and it was not until she was in her early thirties that she was able to study nursing at a hospital near Düsseldorf in Germany.

Upon her return to England, Florence took an administrative post as Superintendent to the newly created Establishment for Gentlewomen in Harley Street, London. Then a few months later, early in 1854, as war erupted in the Crimea, a friend of the Nightingale family, the British Minister of War Sidney Herbert, asked her and her family if she would supervise a team of nurses who were about to leave for Turkey and the war. 'There is no other person in England I know of,' Herbert told Florence's father, 'who would be capable of organising and superintending such a scheme.'

As preparations for Florence's assignment were underway, the war began to escalate. Soon, journalists in London were reporting terrible conditions on the battlefields and in the hospitals at the front. *The Times* was particularly bellicose in its criticism of the government and it fired broadsides at the military planners for their apparent disregard of the health of front-line British soldiers. Its war correspondent William Howard Russell reported with disgust: 'Not only are there not sufficient surgeons . . . not only are there no dressers and nurses . . . There is not even linen to make bandages.'[1]

It was into this chaos that Nightingale and her thirty nurses arrived at Barrack Hospital in Scutari on the eastern side of the Bosphorus in November 1854. At first, the doctors there refused their help and considered the nurses to be little more than an embarrassment. It was only after the first major battle, at Inkerman in Turkey, when thousands of wounded soldiers were transported to Scutari, that the nurses came into their own and the cynical male medics could no longer ignore them or resent them.

Soon after this battle, Florence Nightingale wrote home describing her role as 'General Dealer in socks, shirts, knives and forks, wooden spoons, tin baths, tables and forms, cabbages and carrots, operating tables, towels and soap'. But she was being very modest. An obsession with hygiene combined with her irrepressible drive had already made a phenomenal difference. Throughout the war zone, she managed to slash the death rate caused by infection from 40 to 2 per cent in her hospitals.

The Crimean War brought Florence Nightingale worldwide fame and she was not slow to exploit her celebrity. Her main concern was that the government, the medical establishment and the general public should learn from the dreadful conflict she had lived through and, just as her close contemporaries in the United States would do, she campaigned ceaselessly to bring dramatic reform to both the military medical system and the civilian hospital system. She established nursing schools, raised funds to build hospitals and acted as a government adviser in Britain and abroad. Working when the British Empire was at its peak, she could exert a huge influence over the development of the health systems of many countries throughout the world, an influence still felt today.

Florence Nightingale was a woman of her time and many of her ideas are now considered antiquated and ineffective. It is also particularly ironic that she managed to increase the survival rates of the sick and wounded in spite of being an outspoken opponent of Lister and utterly refused to accept the notion of bacteria, believing instead that infection was spontaneously generated in 'foul air'. Yet, her spirit and many of her ideas concerning the discipline of nursing are as respected today as they were during the 1860s. Most importantly, Florence Nightingale initiated a revolution in nursing that became an unstoppable force.

Today, hospitals and medical care remain far from perfect, but it is important to remember that, at the beginning of the Victorian era, the hospice, the asylum and the doctor's practice were dark, unsanitary places often run by criminals and drunkards. Towards the end of his career, soon after the First World War the American doctor Robert Morris was moved to write in his memoir: 'One of the very greatest changes that I have observed in the past fifty years has been the attitude of the public toward hospitals. Dread of them was general . . . All over the world the very name "hospital" suggested pestilence or insanity; few people would go voluntarily to such a place, no matter how well equipped it was for doing routine work efficiently. Today almost everybody with any illness at all serious wishes to go there.' The most influential people behind this change were Florence Nightingale, Dorothea Dix and Clara Barton.

Much of this transformation came from the lessons of war and through the indomitable spirit of those who experienced the deadly power of infection and disease. But none of the changes seen during the past century and a half could have occurred if the timing had been wrong. The three women at the forefront of this shift in medical practice captured the spirit of reform not only within their field but throughout the entire social structure of the era, a structure that stifled the rights of women and encouraged elitism and sexism.

In the United States immediately after the Civil War, agitation by women for the vote became increasingly vociferous, and in Britain the cause was taken up by the philosopher John Stuart Mill who presented a petition to Parliament calling for inclusion of women's suffrage in the Reform Bill of 1867. The same year, Lydia Becker founded the first women's suffrage committee in Manchester.

But progress was incredibly slow and society was moving towards genuine reform at a snail's pace when another war, far worse than the conflicts in the Crimea or the United States, suddenly accelerated the pace of change. The First World War, the first global military conflict, was all-embracing, and when it was over men could no longer ignore the enormous role played by women, one half of the population which had previously been offered precious little respect and afforded almost no power or influence within society. During the dying months of the First World War American women over thirty won the right to vote and they were joined two years later by the women of Britain.

Once more, war had wrought change far beyond the outcome of political strife and it had altered more than the transitory power structure of fighting nations.

Reconstruction

A social change of equal importance which occurred at about the same time was the way in which those terribly disfigured by war were given the sort of care and consideration they deserved. The art of remodelling a damaged human face is a surprisingly old one. In ancient India the secret of reconstructive surgery was passed from father to son within families of potters who, from as early as 800 BC, had been producing pottery noses for adulterous wives disfigured by vengeful husbands. The secret was so well preserved that the West only learned of the art during the fifteenth century and it took another hundred years before the method gained popularity through the writings of the Italian surgeon Gasparo Tagliacozzi.

In 1794, two doctors serving with the British Army in India witnessed an operation to create a false nose and described their observations in a paper that received considerable attention in London, where the operation was dubbed the 'Hindu method'.

Other accounts of similar procedures have filtered down through history. Perhaps the most startling came from the writings of another Italian, a contemporary of Tagliacozzi, a pioneering medic named Christopher Fiorovanti, who had seen a swordsman slice off his opponent's nose in a duel. Fiorovanti apparently picked up the nose from where it lay in the sand, urinated on it to wash it and sewed it back on. To his amazement, the procedure worked and the nose attached itself to the victim's face without infection.

There are many similar tales. Most have been substantially embellished, while others are certainly entirely fictitious. Indeed, until the twentieth century not only were the technical and medical problems of reconstructive surgery extremely difficult to overcome, but both the military and medical authorities saw little value in making the effort.

A few keen experimenters did make progress in the science of reconstructive surgery and their advances are now seen as having been far ahead of their time. The French surgeon Guillaume Dupuytren, who was born into a peasant family in 1777, rose to become the most esteemed surgeon in France by repairing disfiguring facial wounds using skin grafts and clever stitching. Notorious for his savage tongue and extremely quick temper, he gained a reputation as an eccentric genius because he often operated wearing carpet slippers and a cloth cap.

Dupuytren was an exception. Well into the twentieth century, both French and German military medics were little concerned with developing plastic surgery or reconstructive techniques. The Germans are said to have simply patched up soldiers disfigured in the trenches and sent them back to the front as soon as they could walk. The French showed almost no interest in reconstructive surgery developments made by their British or American allies. One French doctor was reported as saying that 'A plastic surgery patient looked horrible when they went into the operating theatre and ridiculous when they came out.'

Such attitudes seem to have stemmed from a misplaced sense of machismo and the belief that a soldier was not meant to look pretty. As we have seen in the case of nurses, who were accepted as professionals only when the moment was right, the recognition of the benefits of plastic surgery too had to wait for soldiers' and the military's gradual acceptance. The nineteenth-century American doctor John Orlando Roe, who had conducted many experiments in reconstructive technique, wondered poignantly: 'How much valuable talent has been buried from human eyes, lost to the world and society by reason of embarrassment caused by the conscious, or in some cases, unconscious influence of some physical infirmity or deformity or unsightly blemish.'

In Roe's day the most common reaction of doctors to those unfortunates severely disfigured by accident, disease or war was to pack the patient off to a secured unit, which meant the victim simply 'disappeared'. Such attitudes persisted until the 1920s when the earliest technical advances in plastic surgery and reconstructive technique began to make an impact. In that decade much was learned from intensive treatment of the thousands of soldiers who had been severely injured during the Great War. The two most brilliant plastic surgeons of the period were the British doctor Harold Gillies and his assistant, the New Zealander Archibald McIndoe, who had studied at New York's Mayo Clinic for two years after the end of the First World War.

Gillies and McIndoe developed their methods at the British Army base at Aldershot in southern England where they treated more than two thousand soldiers shipped home in tatters after the Battle of the Somme in 1916. These soldiers had been written off by society and immediately institutionalised. Gillies and McIndoe were outraged by such inhuman treatment and set themselves the seemingly impossible task of helping every last soldier in their care. Working eighteen hours a day for almost eight years, by 1928 they had, to some degree at least, treated all two thousand men. During the 1920s, plastic surgery was really an entirely experimental science that required the Aldershot team to make up the rules as they went along. For their part, the human guinea pigs, those brave, discarded soldiers of the Somme, had little to lose.

But even Gillies and McIndoe's heroic efforts went almost totally unnoticed by the medical fraternity. By the outbreak of the Second World War, the team had split up: Gillies became a consultant in plastic surgery to the RAF and McIndoe took a position in a civilian hospital. But after the Battle of Britain in 1940, when some four thousand airmen suffered burns and facial fractures requiring reconstructive surgery, the two men once again pooled their vast experience and performed near-miracles reconstructing the faces and hands of horribly disfigured pilots, sailors and soldiers.

In some extreme cases, however, not even Gillies and McIndoe could do much to repair severely disfigured patients so they had to begin again from scratch. They moulded plastic bones, which they then inserted under mangled flesh, and pulled down flaps of skin from the scalp to cover completely reconstructed noses. Sometimes eye sockets had to be totally reshaped. Completely fabricated chins were attached to surviving, usable bone and false eyes were inserted into empty sockets.

Such procedures could take up to three years to complete. One patient, Flying Officer James Wright DFC, was so badly burned when his Spitfire was shot down that he needed a total of forty-six operations and eight corneal grafts. On another patient, a young man whose hands had been turned into a mass of distorted flesh by a cockpit fire, McIndoe conducted thirty-eight different operations. At times he had to stitch skin grafts the size of postage stamps on to the man's fingers.

Meanwhile, American teams had begun their own experimental surgery on disfigured USAF pilots, soldiers and seamen severely wounded in Europe, North Africa and the Far East. The American Society of Plastic Surgeons had been founded in New York in 1920, but just as Gillies had been ignored in Britain so little notice was taken of the pioneering work of Jacques Maliniac and Gustave Aufricht, leading lights of the ASPS. Only slowly did the medical establishment come to accept that amazing results could be achieved, that the faces of many servicemen could be saved, allowing them to lead lives as nearly normal as possible.

The Korean War provided a further opportunity to develop the science of plastic surgery, and during the mid-1950s surgeons made stunning advances that enabled them to develop new ways to rebuild faces and hands. Methods such as wiring facial fractures, grafting tiny pieces of skin, creating effective epidermal layers that breathe like natural skin, manufacturing lighter and stronger replacement bones and techniques for reconstructing joints once considered revolutionary have now become common-place.

Today, modern materials allow for remarkably sophisticated surgery. 'Smart plastics' that can accurately imitate the properties and characteristics of human skin are used along with ultra-light-weight metals to fashion bone and cartilage; and computer programmes that provide insights into how best to remodel a patient's face or hands have aided the plastic surgeon enormously. Many of these advances have come from fields of research that have little to do with medicine but can be employed by techni-cians and medical physicists to improve surgical technique and postoperative care. As with so much modern technology, devel-opment of these ideas and methods comes from a web of progress, a complex network of technology and the imaginative cross-fertilisation of disparate disciplines.

One recent tragedy serves as an example of how we can learn lessons from violent conflict. When a terrorist bomb exploded in a nightclub in Kuta, Bali, in October 2002, hundreds of victims suffered burns and horrific wounds from the explosion and the subsequent fire. Many of these people have been treated at Royal Perth Hospital in Western Australia by skin-graft specialist Dr Fiona Woods. Working with a small team, Woods has developed a solution called CellSpray which is made from a patient's own healthy skin cells. Sprayed on to the wound, CellSpray creates an epidermal layer that allows the patient's skin to grow over the injury. So far some two thousand patients have been treated using this new technology, and it is believed that another thirty thou-sand patients from around the world could benefit from it directly.

Reconstructive surgery is often confused with cosmetic surgery, a practice which has a rather tarnished image. Too many pop stars and actors going under the knife have given this science a new and unwelcome public image, which has been exaggerated further by the recent taste for grotesque TV shows treating plastic surgery as a form of entertainment. Once, reconstructive surgeons were ignored or scoffed at, their work considered irrelevant; today, celebrity-obsessed society and a media preoccupied with stoking public hysteria have confused the perception of what reconstructive surgery actually does. Yet, ironically perhaps, plastic surgery continues to make a huge impact upon the lives of many civilians who, through birth defects, accidents or disease, have benefited from modern reconstructive technique and the lessons handed down from pioneering war surgeons of the two world wars of the twentieth century.

Facing the Horrors of War

Until the twentieth century only a handful of enlightened individuals considered the impact of war on the mental health of hard-pressed fighting men. For example, in Britain an occupational therapy unit was created in Chatham in 1820, and the earliest recognition that the horrors of battle could disturb the mind came from the Franco-Prussian War. During that conflict, the military historian Captain Fritz Hönig wrote: 'I am not ashamed of owning that the French fire affected my nerves for months after. Troops that have survived an ordeal of this kind are demoralized for a considerable time.'[1]

With this description, Hönig was grasping for a definition of the very real mental and physical affliction we now call 'shell shock', a condition which was then many years away from being recognised by medical science. Almost half a century after Hönig's time, during the First World War, the precise physiological mechanism of how exploding shells affected the brain was unknown, and there was no effective method of treatment. Consequently, many hundreds of sick men on both sides were labelled cowards and deserters and some were even executed by their own commanders. As late as the 1940s, after years of research into trying to find the cause of shell shock, some military leaders still failed or refused to grasp that this disorder and other psychological traumas were genuine illnesses that could not simply be put down to cowardice or cheap tricks to secure a passage home.

Perhaps the most shameful example of such denial was that of the American four-star general, George Patton. During a visit to a field hospital in Nicosia, Cyprus, in 1943, Patton was introduced to a soldier whose diagnosis label pinned to his chest indicated that he was suffering from 'psycho-neurosis anxiety state; moderate/severe'. This was inaccurate; the man was in fact suffering from dysentery and malaria and running a dangerously high fever that had caused him to become disorientated. Patton studied the label with growing contempt and then asked the man what was really wrong with him. Barely lucid, the soldier replied: 'I guess I just can't take it.'

Patton flew into a rage, slapped the man with his gloves and seized him by the shoulders. The patient instantly went limp and started to whimper, which enraged Patton even more. Pushing aside the orderlies and the doctor escorting his party, Patton dragged the man out of the tent and threw him face down into the mud. He then drew his pistol, waved it at the soldier and screamed that he would have the coward shot. It was only the intervention of a senior medic who came upon the scene in the nick of time that saved the soldier's life.

By 1943, Patton had already gained a reputation as mentally unstable himself. A few weeks before the incident in Nicosia he had gone on record at a field hospital as recommending an amputee be left to die because '. . . he is no God-damned use to us any more'.[2] Within hours of Patton's mistreatment of the patient in Nicosia, a journalist stationed in Cyprus heard of the outrage and the sorry tale was splashed across the newspapers of America, damaging the general's career irreparably.

Patton's was an extreme case, and, for the most part, between the wars the military establishment and the medical fraternity gradually became aware that mental stress during war was a tangible and treatable problem that needed to be addressed. Slowly, the lessons learned from shell-shock victims of the First World

War generated a more enlightened approach to the mental stability of fighting men. By the conflict of 1939–45, the British Army no longer executed men for desertion or cowardice. In fact, the very word 'cowardice' was no longer used officially to describe men made incapable of fighting through psychological disorders.

Soldiers who suffer shell shock usually follow a succession of recognised and documented symptoms. They become disorientated, irrational, often swinging from moods of extreme violence to complete apathy; their hands shake and their eyes become unfocused. To the commanders in wars prior to those of the twentieth century, such reactions could easily have been mistaken for a form of insanity or cowardice and so soldiers may have been discarded as beyond hope or executed for dishonouring their fellows. Only as medical science developed and doctors acquired a clearer understanding of certain brain functions was the real mechanism of shell shock revealed.

The physical damage associated with shell shock comes from the effects of projectiles exploding close to a soldier who might then be wounded by shrapnel and buried under mounds of earth and mud thrown up by the explosion. Brain damage may occur thanks to an elaborate series of events.

As the shell explodes, it creates a temporary vacuum around the impact site. This vacuum is almost immediately filled by an inrush of air, which causes a very rapid increase in atmospheric pressure in the region. This sudden alteration in pressure can have a dramatic effect upon the highly sensitive tissues surrounding the brain, causing often serious debilitation.

A very different problem is the effect of disturbing experiences on the mental state of a soldier. For many years military analysts and doctors believed that fighting men 'grew used to combat'; but this is simply not true. For a soldier on the field of Agincourt, fighting in the second Gulf War or in any conflict in history, the horrors of military engagement are manifold and constant. Human beings do not adapt to these experiences; they merely suppress the traumas, and store up the anguish they cause.

The psychological damage caused by fighting takes many forms and has many different sources. One pioneering doctor, the US Army officer Captain Frank Hanson, made a detailed study of battle trauma in August 1942 when he volunteered for front-line action specifically to gather data. His striking conclusion was that many cases of psychological disturbance in soldiers were due simply to lack of sleep. As a result of his findings, the 48th Surgical Hospital of the US Army conducted a series of sleep-deprivation experiments involving two hundred patients and found that a mere thirty-six hours of almost uninterrupted sleep cured 30 per cent of them. Of those left to sleep for forty-eight hours, 70 per cent were able to return to battle within a week.

Naturally, lack of sleep is only one of many causes of trauma. Some soldiers, sailors and pilots are pushed to their physical and mental limits for so long their brains simply 'shut down' to allow them to rest from the experiences forced upon them. Others are merely reacting to the immense physical exhaustion of continuous duty in the most uncomfortable environments imaginable.

As examples of humans living *in extremis*, the plight of the First World War soldier fascinated psychologists at home. Freud was particularly interested in the deep-rooted causes of psychological disturbances witnessed in many thousands of men returning from the front. In his *Introductory Lectures in Psychoanalysis* (1917), Freud attempted to address what he saw as the psychogenesis of shell shock. He incorrectly ascribed the symptoms to what he called an 'hysterical response' and used the examples of shell-shocked soldiers in an attempt to explain the 'puzzling transition from mental to physical symptoms'. In other words, Freud concluded that a mental disturbance could cause recognisable and even treatable physical effects, but he was mystified as to the mechanism involved. He never did succeed in finding that mechanism and he was entirely wrong about the causes of shell shock. That said, he should be credited for being one of the first to recognise that there must be underlying patterns that generate psychosomatic effects.

Immediately after the First World War the new science of psychotherapy became more widely practised. The fundamentals had been pioneered before the war by Freud, Jung, Meyer, Ernest Jones and a cadre of radical researchers, but now, as one historian has put it: 'Perhaps shocked into modernity by the horrors of the war, the lay public became increasingly receptive to the notions of psychotherapy.' Indeed, the central tenets of the new science, often misinterpreted and confused by oversimplification, frequently made the pages of the national newspapers of the day. 'Freud's books,' grumbled the *Saturday Review*, '. . . are now discussed over soup with the latest play or novel.'

Yet such publicity has gradually made an impact. The many mental sufferings of soldiers through successive wars have supplied psychologists with knowledge they would never have otherwise acquired, knowledge that has been successfully employed to help those caught up in civilian tragedies that mirror the terrible conditions of war.

During the 1980s two new terms were coined to describe particular psychological conditions: 'post-traumatic stress disorder' and 'survivor guilt'. Both terms derived from the language of battlefield psychology but are now used to describe the condition of people who, in civilian life, have suffered traumatic experiences: victims of major accidents which involve many fatalities, survivors of terrorist attacks, victims of mugging or rape. Many unfortunate people, however, suffered from these conditions long before they were recognised as genuine reactions to shocking experiences.

Until the 1960s very little was known of the enormously complex physio-psychological protective mechanisms of the human brain. During less enlightened times, when shell-shocked soldiers were routinely branded 'cowards', anyone who seemed severely disturbed by a traumatic experience was merely told to 'pull themselves together'. Thankfully, today, the police, the medical community and psychologists all work together to alleviate the symptoms of post-traumatic stress disorder and survivor guilt, now known to be potentially debilitating effects of trauma.

The knowledge psychologists have gained from the psychological trauma of servicemen has been of constant help to their

profession and has often helped clinically detached military commanders to understand their own men better. It has led to better
management of the stress and trauma of battle but the civilian is
also a beneficiary of these changes because, in addition to helping
the theoreticians develop more refined interpretations of human
behaviour in extreme situations, the research conclusions of the
war psychologists have fed back into the practical work of trauma-
management specialists. Together, these developments have
improved the quality of life of many people whose psychiatric
problems might otherwise have remained totally undiagnosed,
ignored or completely misunderstood.

Medical Organisation

Within both the theatre of war and civilian life the growing knowledge base of doctors and continuous improvements in technology lose almost all their effect if they are not supported by a well-designed infrastructure, efficient planning and professional organisation.

In ancient times there was almost no planning for the medical requirements of the soldier. Kings and military leaders gathered often ramshackle armies and went to war with no thought for the medical needs of their men or the benefits that could be gained from deploying a healthy, well-fed force. These societies went to war with the incantations of priestesses ringing in their ears; if they fell, they simply prayed.

The Romans, ever the innovators, were probably the first to construct field hospitals, which they called *valetudinaria*. And, although they were effectively little more than tents in which the sick and wounded were given time to recover or die, they represented an idea far ahead of its time. When the Roman Empire crumbled, the notion of the *valetudinarium* went with it and only in the East, where some knowledge of science, mathematics and medicine was preserved (and sometimes improved upon), did military medicine survive at all.

As we have seen, many innovations within medicine have resulted from shifts in the attitude of governments and the highest echelons of the military (who in the modern era have been propelled into action through public accountability). The organisation of military medicine into an effective institution is no exception to this principle.

The ninth-century AD Byzantine army of Leo VI is said to have included surgeons and primitive field hospitals, as well as a form of ambulance corps made up of stretcher bearers, known as *deputati*, who carried the wounded from the field of battle. But when we next hear of organised medical recruits in Europe it is on the field of Agincourt, half a millennium later. In October 1415 a group of twenty surgeons, often labouring up to their elbows in blood and viscera, served Henry V's 32,000-strong army.

The first field hospitals appeared as late as the Napoleonic Wars at the beginning of the nineteenth century. The French created what they called *hôpital ambulant*, a term which led to the name 'ambulance'. Litter-bearers were employed to carry the wounded from the battlefield to the field hospital where the surgeons sliced and sewed. But this system was poorly organised, underfunded and considered largely irrelevant by most commanders. As a consequence, it was to take the influence of public opinion during the American Civil War and the Crimean War in Europe fifty years later to effect genuine change. And once begun, that change was rapid.

At the onset of the American Civil War in 1861, the soldier's lot was little different from that of any other throughout the long, torturous history of war. If soldiers fell in battle they would be left to the care of their comrades, while a lucky few might be dragged to a tent behind the lines where an overworked and undertrained doctor would do his best. By the end of the war, only four years later, hand-wheeled litters were in use, specially trained ambulance men whose sole task was to ferry wounded from the field were at work, and the makeshift tents of 1861 had been replaced by field hospitals with operating tables, anaesthetics, bandages and water supplies.

In the Crimea, the changes instigated by Florence Nightingale highlighted the scandalous conditions prevalent at the start of the war. The British public had been shocked to hear how their boys were treated by their own officers. Eventually, pressure from British civilians forced the highest ranks within the army to listen and to initiate change.

News of these developments and the improvements to the efficiency of armies that resulted spread rapidly. By the time of the next major European conflict, the Franco-Prussian War of 1870–1, a Swiss invention, the *mandil di socorro*, a form of foldaway carrying apron, or folding stretcher, was in common use. This, and seemingly obvious improvements, such as an awareness that wounded soldiers should be grouped into urgent, non-urgent and minor categories, saved thousands of lives.

By the end of the nineteenth century, the medical services of Western armies had reached a level of sophistication that would have been quite unrecognisable to earlier generations. An order of battle dated 1880 shows that a British Army force of 36,000 men had the use of 14 field hospitals, 8 stationary hospitals and 2 general hospitals. Well-defined lines of communication between the front and the medical support units had been arranged, detailed projections for the expected number of casualties had been calculated and appropriate resources made available before engagement.

A few decades later, in 1917, Ernest Hemingway, then a volunteer in the American Red Cross, witnessed the chaos and squalor of war in Italy and wrote vivid accounts of makeshift hospitals in old huts, apparently endless shortages of medicines and stories of soldiers dying unnecessarily. Yet, compared with any earlier conflict, the level of medical services organisation during the First World War was high. Motorised ambulances were used for the first time, trains provided links behind the lines and medical knowledge, especially that of infection management, had greatly improved.

By the Second World War, a medical service for the army, navy and air force was in operation that would be recognised by a serviceman of the twenty-first century as essentially similar to their own. A military campaign document of 1940 – an updated version of the one produced by the British Army in 1880 – gives details of how trains, ships and planes served to augment motorised ambulances driven by specially trained teams of medics. In this plan, three large field hospitals liaised with a recent innovation – the specialised hospital. These are listed as: field dressing stations, advanced surgical centres, rest stations and convalescence centres outside the theatre of war.

And, of course, new and better ways of running the medical services of the armed forces went far beyond the battlefield. The lessons the military learned for organising its personnel and coordinating the treatment of sick and wounded soldiers flowed constantly into civilian life. The very doctors and surgeons who improved the military system returned home to initiate similar plans in civilian hospitals. They chaired committees that altered the way health services were organised and managed and they wrote influential papers and books to spark public interest and push forward developments in health management.

However, new military thinking often took time to produce valuable change and sometimes it took a tragedy to shake the complacent old school into accelerating improvements. One such tragedy occurred in the early 1950s in Harrow, in north-west London.

At 7.30 on the morning of 8 October 1952 a commuter train carrying several hundred people and travelling at 60 mph ploughed into two locomotives that had collided only seconds earlier directly alongside the platform of Harrow & Wealdstone Station. Huge pieces of metal shot into the air and whole sections of the mangled trains flew along a crowded platform, causing terrible devastation.

Immediately, rescue crews and medical services from across London raced to the scene. First to arrive was a fleet of ambulances from the local hospital in Harrow. A few minutes later, a dozen military ambulances arrived from a US Air Force base a few miles away. Both teams were manned by dedicated crews who worked hard to save those hurt and to help rescuers extricate the dead and severely injured from the wreckage. But of those passengers taken to hospital by the Harrow ambulance teams, only ten out of ninety-five of the injured survived. Meanwhile, the Americans, who had only that month deployed the first paramedic teams in the world (in Korea) were able to treat the accident victims at the scene rather than spending valuable time transporting them to hospital before medical attention could be administered. Strikingly, every patient treated by the air force medics that day survived their injuries.

Today, that 'golden hour', the first sixty minutes after an accident occurs, is considered paramount to the survival chances of anyone involved in a serious accident. This simple truth was not grasped until the end of the Second World War, but young American medics returning from the war in Europe were quick to extol the virtues of creating mobile forces trained to give life-saving treatment *in situ*. Training programmes were quickly established to teach advanced first aid techniques that had also been rationalised and refined on the battlefield. From these programmes came the concept of the 'paramedic', originally a corps of medically trained personnel who parachuted into the scene of an emergency. Soon, civilian hospitals, first in the United States and later in other parts of the world, developed their own quick-response teams for dealing with civilian accidents and emergencies. And, although few of these specially trained individuals were able to parachute (or indeed needed to), the term 'paramedic' stuck. Indeed, the notion of paramedical services is now fairly common throughout the world. One of the best examples is China's 'barefoot doctors', paramedics who operate primarily in rural districts. In Russia, paramedics who work in the countryside are called *feldsher*.

In 1946, less than twelve months after the Second World War had ended, the British government passed a bill that established the National Health Service, a system in which the payment of a national insurance contribution deducted automatically from an individual's pay cheque provided a system of 'free' health care for all. This change was in part due to the election of a new and politically radical Labour government in 1945, but it was also a shift prompted directly by lessons learned during the war.

Doctors returning from the war were keen to make changes in health care. As one historian has written: 'Wartime medicine gave doctors a vision and a voice.' But there were other forces at work. At the outbreak of war in 1939, the British medical system was in a mess, disorganised and with no single voice. The Ministry of Health took over the system and created a network of 1500 public hospitals and more than 1000 private institutions that performed their wartime duties with unprecedented efficiency and economy. After the Second World War, the government was quick to appreciate that this emergency plan worked so much better than the *ad hoc* methods of the prewar years and that it should become the new template for postwar health care.

War has constantly engendered medical innovation. Sometimes this innovation has been serendipitous, at other times it has been the ceaseless, dedicated probing of rare individuals that has produced crucial discoveries and stunning breakthroughs. And yet, as this account shows, equally remarkable has been the depressingly slow pace of change in perception and the acceptance of new, sometimes radical, ideas.

Of course, our attitudes, systems and methods are far from perfect. In Great Britain the National Health Service remains a rare and wonderful thing, but it is also deeply flawed and becoming increasingly difficult to manage. At the same time, the system employed in the United States and by many other Western countries is often seen as too dependent upon profit margins, so that the plight of the sick becomes secondary to fiscal concerns.

These criticisms may be true, but if we need any further evidence that change, although slow, has given us a better world (in large part thanks to the lessons of war) we need only compare the extent and complexity of modern health care systems with the hospices of Renaissance Rome, Paris or London; or conjure up the vision of an injured soldier lying on a rain-drenched battlefield, left behind to die as the army moves on.

PART 2

FROM

THE ARROW

TO

NUCLEAR POWER

In Part 1 we saw how medical advances have come from the battlefield, how healing has derived from the infliction of injury. Part 2 will consider the weapons, the very devices that cause injury in the first place, and how their development has altered radically the technological and social basis of our lives.

This change has been truly radical. The human insistence upon deadlier arms played a key role in sparking the Industrial Revolution and it was crucial to the evolution of the technological society. The technology of war has been of central importance to the development of the production line, the factory and standardisation. It has led to miniaturisation, which created extensive developments in electronics, from which computing has grown. The chemical industry arose out of the need for manufacturing explosives. In addition to these technological advances, the development of firearms led to the first international industry – armament sales – and the sociological and political consequences of this advance cannot be overestimated.

It has been said that the gun is the most important invention in the history of mankind, but in order to appreciate the importance of weapons development in the evolution of technology, and indeed its social affects, we must go back to the beginning, to the earliest weapons, devices that man invented long before the gun.

Sticks and Stones, Bows and Arrows

Stone Age man fashioned tools out of flint, horn and bone, but he used the same skills to design and shape weapons with which he could hunt, defend the tribe and attack neighbours. These were the first weapons – stones to throw and sharpened bone and flint to use as the most basic projectiles, daggers and clubs.

The concept of the arms race is deeply rooted in the development of civilisation and it first became apparent in the most primitive cultures. Innovative behaviour and taking advantage of good fortune led to a tribe's dominance over rivals. Innovation came from the ability to make the most of what was available. Luck came down to resources. In the earliest cultures, tribes were evenly matched because, for the most part, resources such as stones and bones were universally available. Later, however, when metals were first used and before trade spread these metals, the tribe with easiest access to such rare and valuable materials quickly gained superiority.

One of the earliest technologies is that of metalworking, an example of a craft that ultimately developed into a huge industry thanks to the joint uses of metal – the making of weapons and the fashioning of tools. The lessons gained from one use were, of course, applied to the other.

It began with copper when, by accident, Stone Age people of about 10,000 BC discovered various uses for the metal. Rich seams of pure copper are rare. More usually the metal is found as an ore from which the pure metal may be extracted. The process via which this was first achieved presented challenges, the meeting of which led to important and widely applicable developments.

The extraction of copper required the union of two quite separate skills. The first was the production of heat far greater than an ordinary cooking fire could provide. The temperature of a typical camp fire is about 700°C, but copper does not even soften until it reaches 800°C and 1000°C is required to liquefy the metal so that it can be shaped and moulded. The second skill was the removal of impurities from copper ores such as the greeny-blue malachite which contains carbonates, or the ore, chalcocite, which has a high sulphur content. Rather unsophisticated weapons could have been fashioned if, by chance, a hot flame softened the metal so that it could be shaped to some degree, but in order to use copper to its full potential new techniques had to be developed.

To achieve a hotter flame, a fire needs to be forced; that is, oxygen must be pumped into the base of the fire, for example by blowing air through a narrow pipe that has been coated in clay to protect it from the heat. Once this technique was developed the next step was to enclose the fire in a primitive furnace so that the heat could be better controlled. In this way temperatures of up to 2000°C were obtained. Once the ore had been baked in this way, the metalworker was left with copper oxides. In the second stage of the extraction process, the smith had to reduce the amount of oxygen reaching the ore and the fire, so as to drive the oxygen from the oxides.

In the final stage of the extraction process, an array of chemicals was used to wash away stubborn impurities and contaminants. The oldest evidence of a tribe capable of copper extraction comes from a site in Tal-I-Iblis in modern-day Iran where metal weapons and clay pots stained with residues from the copper-refining process have been found and dated to about 5000 BC.

The skills required to make these artefacts and the lessons learned by these primitive peoples, most likely through trial and error, were later applied to the chemistry laboratory and gave the early metal craftsmen a certain mystique. Knowledge of how to control the temperature of a fire was at first an almost occult power but it was gradually seen as simply the application of intellect and skill, and the furnace became a core part of the technology for any civilised people. From the purification of metals came the alchemist's art, a knowledge of how to use acidic and alkaline solutions, how to remove chemical groups (such as sulphates or oxides) and how to substitute them if required.

It was almost certainly again through luck and trial and error that the next great leap in metalworking took place. Copper was a highly prized commodity and deemed beautiful by primitive man, but because it was so soft it was less than satisfactory for the weapon maker. Some seven thousand years ago it was found that if 10 to 15 per cent of the copper is replaced by tin, the result, bronze, is both harder and stronger.

Bronze was the first man-made alloy and its discovery altered the manufacture of weapons radically. It enabled smiths to fashion a variety of different swords and daggers and it was found that a handle made from metal and covered in sinew or skin not only gave a far better grip but allowed more skilful manipulation of the weapon. Bronze could be shaped as easily as copper and this enabled the improvement of the spear. Although spearheads had been made with sharpened stone or flint, bronze heads were lighter and could be shaped more easily. Furthermore, whilst stone or flint heads had to be bound to the shaft, with a bronze head the shaft could be inserted into the head once a hole had been bored to accommodate it.

Bronze was also decorative and it quickly superseded copper for ornamentation. Around the same time, 4–5000 BC, another alloy, brass, made its first appearance. Brass is a blend of copper and zinc. Although brass was not as good as bronze for making weapons it found many domestic uses.

Because tin was needed to make bronze, people who had settled close to the sites of tin deposits prospered, either because they could manufacture better weapons than their neighbours or because they could become wealthy through trade. Tin was one of the most useful resources found by the Romans in ancient Britain (in Cornwall in particular) and although by this time Roman legions no longer used bronze weapons, the smithies of the empire found a multitude of uses for tin.

Bronze and copper were ornamental as well as useful to primitive weapon makers but the same could not be said for iron, a metal that is rather ugly and dull. The introduction of iron, which was first extracted and used around 1500 BC, was an even more important change to technology and society than copper or bronze. Iron is a great deal stronger than bronze, it is more plentiful, and, although it rusts, it is far more versatile and can be fashioned into a broader range of weapons and tools.

Putting a precise date on the arrival of the Iron Age is rather meaningless because it did not happen simultaneously in different parts of the world. Different cultures first utilised iron and followed their own paths of development in using it at different times depending upon their resources and needs. However, knowledge of this new technology – the ability to fashion this hard, versatile metal – spread quickly from one region to another because, apart from the many uses of iron within civil societies, different military requirements drove change and fuelled progress and innovation.

The Greeks built sophisticated iron foundries and produced high-quality weapons, including a leaf-shaped sword called the *akinakes* and the curved *kopis*. Iron is more difficult to extract than copper and so it requires better smelting techniques, especially during the crucial stages in which the control of oxygen in the furnace is essential. Just as copper alloys had been created by blending copper with other metals, it was only a matter of time before metalworkers devised iron alloys that had the properties required by the military.

The first people to exploit these methods were the Vikings, who knew how to carbonise iron to make it much stronger. Carbonisation enabled them to make larger swords than those used by the Greeks and the Romans. Viking warriors therefore had a longer reach and, until the secret of carbonising iron leaked out of Scandinavia, this simple technological advantage made them militarily stronger.

The knowledge and experience they gained from fashioning superior swords and daggers enabled smiths also to make stronger horseshoes, better cooking utensils and more robust and longer-lasting tools. They devised techniques that greatly strengthened the walls of those castles and palaces that used metal struts and supports. From its first use as a material for making weapons, iron became indispensable to civilised societies everywhere. In short, it radically altered civilisation.

In tandem with the evolution of the sword, the spear and the dagger came the development of the bow and arrow, a weapon mentioned in the Old Testament and employed to great effect by the armies of the ancient Greeks. The level of skill required to make a bow and arrow is far greater than that needed by primitive men to fashion simple side arms. Wood of suitable flexibility is required, so the bow evolved gradually from a rather primitive, short-range weapon to one of remarkable precision and power.

The simple bow used since ancient times was replaced by the composite reflex bow, a far more sophisticated weapon made from a variety of different woods and employed from about 1500 BC. The final step in the evolution of the bow came about a thousand years later with the invention of the crossbow, a radically different and far more deadly weapon than its predecessors. First devised by the Chinese around 500 BC and later used extensively by the Romans, the skills required to make an efficient crossbow – precision construction methods and a sound knowledge of practical mechanics, metalworking, and woodworking – were then lost to the West when the Roman Empire collapsed and were only rediscovered during the Middle Ages.

Evidence that the Romans used the crossbow comes from a passage written by the historian Vegetius in his *De Re Militari* which he dedicated to the Emperor Valentinian in AD 385. However, the crossbow only reached its full potential when it was introduced into the armouries of Europe during the tenth century. For the soldiers of the time the crossbow was considered a 'super weapon' and it proved to be a startling development in the arms race of the day, a weapon widely feared, for, in the hands of skilful cross-bowmen, it could pierce chain mail from a distance of at least three hundred yards. It provided an astonishing new level of power and coming under attack from a crossbow-wielding enemy was referred to as 'a bolt out of the blue'. Indeed, the crossbow was such a devastating weapon that there were calls for it to be banned. In 1097, Pope Urban II outlawed the use of the crossbow, and, four decades later, Pope Innocent II convened a Lateran Council in which it was agreed that use of the weapon should be prohibited '. . . under penalty of anathema', calling it 'the dastard's weapon'. Soon after this the Holy Roman Emperor Conrad III banned the use of the crossbow throughout his realm.

This decision (which was completely ignored by most military leaders of the age) had little to do with any moral prerogative and everything to do with the threat to the *status quo* presented by this new, deadly weapon. In the hands of a peasant a crossbow could kill a knight in full armour, and the knight was, in the heyday of the chivalric tradition, ordained by God to fight. The crossbow had the potential to become a social leveller and it was therefore quickly perceived as a danger to those in power.

Efforts to ban the crossbow failed because the need for military superiority was a stronger force than almost any other sociological or cultural imperative. For all their fine words and threats neither Pope Innocent II nor Conrad III could accept the notion of unilateral disarmament, and so the crossbow enjoyed almost half a millennium as one of the most feared and deadly weapons in the armoury, reigning supreme until the early fifteenth-century when it was superseded by the gun.

Guns, Steam and Revolution

The need to outdo one's neighbours and to acquire their latest technological advances illustrates one form of arms race. The other form it may take is in the interplay between the development of weapons and the creation of defensive systems, processes that are two sides of the same coin and which share a symbiosis.

As communities grew so too did their defensive systems, the forts and castles designed to protect those communities. These defensive systems grew larger and stronger because the weaponry arrayed against them in the arsenal of the aggressor evolved with ferocious speed.

The earliest recorded fortifications were in Egypt. Two thousand years later, the Greeks also experimented with defensive systems but their cities remained surprisingly vulnerable to attack and the only genuinely fortified part of any Greek town or settlement was the *akropolis*, usually to be found on a hill at the centre of a town. But then, as with many things, the knowledge required to construct these strong stone forts and castles was lost, disappearing with the Roman Empire only to be rediscovered in about AD 1000.

To counterbalance the advances in fort construction the Greeks also became the first to build siege engines. The first known catapult-style siege engine was designed by Archimedes in the third century BC in Syracuse and the Greeks took the design of rams and catapults to a level of great sophistication.

Many lessons were learned from this interaction between increasingly elaborate attack weapons and the ingenuity of architects and builders. For the attacker, the crucial qualities for their devices were manoeuvrability, power and durability. This meant that the designers were always looking for new ways to build and improve the mechanism for the delivery of projectiles while making their machines stronger and more destructive. As we saw in the Introduction, Leonardo da Vinci, wearing his military engineer's cap, became a master of siege-engine design and promised great things with his plans for catapults, mortars and slings, although, like most of his ideas, these plans probably never went beyond sketches in his notebooks.

Another aspect of the aggressors' art was knowing how to make the most of their devastating machines. The Greeks pioneered some of the mathematical techniques needed to calculate the trajectory of projectiles and they also knew how to judge the power and efficiency of an engine according to its size. It is no mere coincidence that the Greeks, who were deeply enamoured of mathematics, were also among the most skilled designers of large military equipment, and that Archimedes, the man credited with the construction of the first military catapult, was also an acclaimed mathematician.

The skills acquired during this early arms race found plentiful application in civilian life. The lessons gained from the construction of forts – how best to build strong, durable walls, how best to transport large blocks of stone and to move them into place, how best to construct staircases – all of this and more became part of the civilian architect's craft. In the same way, the engineering knowledge gained from designing levers and pulleys, transportation systems and siege engines that could be taken apart and rebuilt at the scene of battle fed into other areas, helping in particular the civil engineer.

Leonardo, writing of his siege engines towards the end of the fifteenth century, lived on the cusp of a new age in weaponry and defensive systems, and as well as boasting of the way he could construct slings and catapults he also referred to the very latest technology of the age, the use of cannon.

The precise date and place the cannon was first used is open to question. A surviving note in the Florentine government archives dating from 1326 details a business transaction involving payment for metal bullets, arrows and cannon for the defence of the city. In the same year the historian Walter de Milemete wrote a document for King Edward III to which he added a drawing of a knight lighting a vase-shaped cannon. Interestingly, the projectile in this depiction is a single sturdy arrow which lends support to the idea that the earliest cannon were simply barrels of gunpowder used to hurl oversized arrows and rocks, much like a catapult-style siege engine.

De Milemete's drawing is neither captioned nor accompanied by any text, but the first written account of the use of cannon comes from a document that appeared thirty years after Milemete's in which the author describes the attack launched by a French army under the command of King John II against an English garrison at Breteuil in Normandy. The text contains references to 'jets of fire' and 'heavy bolts' but it does not give a very thorough description of the device itself. This first use of the cannon at Breteuil produced mixed results: the French succeeded in capturing the garrison but it was burned to the ground in the process.

The earliest cannon were almost as dangerous for the army using them as they were for those under attack. The materials used to produce gunpowder had to be blended with precision and a stray spark could be lethal. Guns often split open, backfired or simply exploded when fired. James II of Scotland lost his life at the siege of Roxburgh in 1460 when a cannon exploded, and the mortality rate for gunners was one of the highest in the army. Even as late as 1844 the US Secretary of State Abel Upshur, visiting the naval vessel *Princeton*, was killed when a cannon exploded.

Yet for all the problems associated with the cannon, it was considered a fearsome instrument of war. Machiavelli tells us in *The Prince* that during the French invasion of Italy in 1494 many towns fell 'chalk in hand', which meant that it simply required a French officer to place a mark on the gate of a fortress to signify that it was the next target of the King's guns and the defenders surrendered, terrified.

For a while the cannon became the nuclear weapon of its day; the merest suggestion that it would be used was enough to act as a deterrent, and from the moment of its first introduction on to the battlefield it played a key role in the power struggles of the age, influencing the political flux of nations. However, if its significance to military men was huge, the cannon was to play just as great a role in pushing technology forward; it is no exaggeration to say that without it the Industrial Revolution and the age of steam might well have been delayed by some years.

The earliest cannon were constructed from wood but these were, of course, unreliable and dangerous. Soon metal bands or girdles were being fitted to strengthen the barrel but this too proved far from perfect. It was very quickly realised that for a cannon to be safe and for it to last through a military campaign it would have to be made of metal. So, just as the early weapons makers of the Bronze Age relied upon a ready supply of metal, the military designers and engineers who built cannon during the early sixteenth century also needed large quantities of metal.

Throughout Europe iron was used to make cannon, but in England the armies of Henry VIII preferred bronze. In both cases this meant that metals had to be mined and smelted. The Europeans plundered iron deposits and in England copper and tin ores were mined and the metal extracted and purified at foundries established especially to supply the growing demands of armies. But with the discovery of coke, which is formed when the more volatile components of coal are driven off by heat, the rate of improvement in foundry efficiency began to accelerate.

The first person to realise the potential of coke was an Englishman named Abraham Darby, who constructed his first furnace as early as 1709. However, his device was not very efficient and his efforts were almost completely ignored. Indeed, simple economics left Darby's ideas in the dark for almost half a century and it was not until 1783 when another English inventor, Henry Cort, created what he called a 'coke-fired reverberatory furnace' that the process became viable. After designing and building a prototype, Cort managed to produce full-scale furnaces and made a fortune by having the commercial sense to approach the British government with his work and convince them that, with his new machine, he could produce cannon far more cheaply than by the conventional method.

Cort's design for a coke reverberatory furnace set the pace for the improvement and spread of new furnaces. Desolate regions of the North of England and remote parts of Scotland and Wales became sites for their construction and around them communities sprang up and thrived. As the Industrial Revolution took hold, new industries appeared near the furnaces using the metal extracted and purified there.

All of this came from the fact that the proto-industrialists and financiers who took the risk of setting up these complexes only did so because they were given government contracts to supply gunmetal to the armed forces. It has been estimated that between 1793 and 1815, a period during which the Napoleonic Wars had a major influence upon the states of Europe, more than half of the iron coming from the furnaces of England went to make guns.[1] As William H. McNeill has put it:

> Both the absolute volume of production and the mix of production that came from British factories and forges, 1793–1815, were profoundly affected by government expenditure for war purposes. In particular, government demand created a precocious iron industry, with a capacity in excess of peacetime need. But it also created the condition for future growth by giving British ironmasters extraordinary incentives for finding new uses for the cheaper product their new, large-scale furnaces were able to turn out.[2]

In a very similar way it was a war that made the mass production of steel a viable process. Weapon-makers from the third century BC had known that by introducing a small and controlled quantity of carbon into molten iron, the product, steel, was far stronger and more workable than iron. However, making steel was not easy and the costs involved in its manufacture were prohibitive. The traditional technique involved heating alternate layers of coke and iron with a layer of sand (the source of carbon) but this produced a poor yield.

Yet, because steel was so prized, small quantities were manufactured even though the process was expensive and labour-intensive. During the thirteenth century, crossbow steel was used to strengthen the main struts of the weapon, but because this added enormous cost to the manufacture of the crossbow such weapons were only ever made in very small numbers.

Soon after the Crimean War, both English inventors and military planners came to realise fully the potential of technological innovation for the improvement of weapons and a steady stream of inventions came into the British Patent Office. One of the most hard-working and clever inventors of the period was Henry Bessemer. He had devised, among other things, a machine for extracting juice from sugar cane, a furnace for making sheet glass and a ventilating machine for mines. In 1854, at the outset of the Crimean War, he had developed a new form of mortar shell that could travel further and deliver a more powerful blast than the conventional type. In October of that year he approached the War Ministry with his invention but his ideas fell on deaf ears. Disheartened, a month later he happened to mention his work over dinner with a French war minister in Paris and was delighted to receive not only a positive response but the promise of development money.

Returning to England, Bessemer worked hard to refine the design of his new shell, but he soon hit a major stumbling block. He could produce the mortar shell cheaply and quickly (two of the most important requirements for any form of ammunition) but he discovered that conventional mortars then in use could not withstand the force produced by his shells when they were fired. Determined not to give up, Bessemer quickly discovered that the weak link in the chain lay with the iron used to construct the barrel of the mortar. He also knew that steel, which would be strong enough, was prohibitively expensive. The only way forward was to find a way to make steel cheaply.

The story goes that Bessemer was lying ill in bed one day when the method for cheap steel manufacture suddenly came to him. In conventional iron smelting, air is blown into the furnace from the side. If instead air was introduced from the bottom, Bessemer reasoned, the iron could be heated to a greater temperature and steel produced more economically. (Blowing in air from the side created amorphous lumps of material that hindered the process.)

Bessemer's first furnace was built in Baxter House, St Pancras, London. It consisted of a cylinder about four feet high with six horizontal air pipes around the bottom. For the first trial he put in about three hundred kilograms of pig iron and used air with a pressure of twenty psi. 'All went on quietly for about ten minutes,' Bessemer reported in his journal. 'Then followed a succession of mild explosions, throwing molten slags and splashes of metal high up into the air, the apparatus becoming like a volcano in a state of active eruption. No one could approach the converter to turn off the blast, and some low flat zinc-covered roofs close at hand were in danger of being set on fire by the shower of red-hot matter falling on them.'[3]

Later attempts were more successful and by 1856 Bessemer's furnace was producing high-grade steel at an economic price. During the next fourteen years he accumulated a personal fortune of more than £1 million from his system. He was knighted in 1879.

These technological advances offered great potential but a furnace could only churn out metal for industry if the ore was there to be purified, alloyed or smelted, and initially at least the mining industry was struggling to keep up. As the demand for metal increased, the mines were dug deeper to gain the most from a deposit, but before long these mines became unusable when they flooded with water. Furthermore, the deeper the mines were sunk the harder it was for the miners to breathe and the longer it took for the metal to be brought to the surface.

The seventeenth-century English writer and social commentator Celia Fiennes, who travelled widely and wrote detailed accounts of her adventures around Europe, visited the West Country and saw the mines for herself. She left us a vivid description of the efforts that went into mining metals used for gunmetal, and the technical problems facing the mining companies. At Redruth in Cornwall in 1695 she saw: '. . . a hundred mines, some of which were at work, others that were lost by waters overwhelming them. They even work on the Lord's day to keep the mines drained,' she reported of the miners. 'One thousand men and boys working on the drainage of twenty mines.'[4]

The problem of how to mine deeper and how best to exploit deposits so as to meet the pressing and ever-increasing needs of the military was one that occupied the thoughts of engineers across Europe for the best part of two centuries. It was a conundrum during an age of almost continuous military conflict in one form or another and at a time when the major powers in Europe, primarily the British and the French, were dedicated to empire building. No one working on ways to extract more metal more quickly could have imagined it, but when the solution to this problem was found it led directly to the invention of the steam engine and, with it, the modern technological era.

A number of developments gradually started to make life in the mines a little easier for miners and at the same time increased efficiency. From experiments linked to the need for a water pump, the existence of the vacuum was discovered during the first decade of the seventeenth century, and, in 1654, a German engineer named Otto von Guericke created the first vacuum pump. His ideas were taken up by Denis Papin, a French Huguenot refugee living in Germany who constructed a cylinder containing a piston which could be moved by steam produced from boiling water at the base of the cylinder. When the water cooled, the piston moved back down because of the vacuum that had been produced beneath it, and this was used to lift a weight with a pulley.

Papin's creation was still some considerable way from a steam engine or even a simple pump that could be used in the mines.

However, it formed the basis of a steam-driven device built by the Englishman Thomas Savery that could do practical work. Savery's machine was made from a set of valves and worked on a similar principle to Papin's demonstration model: water was boiled, the steam pushed a block and water could then be drawn into the vacuum that had been created. It was in effect a 'steam-powered vacuum cleaner'. Savery's device proved popular with mine owners and it was used for almost a century after its invention in 1695. Known as the 'Miner's Friend', it could generate up to twenty horsepower and was in use in more 150 mines as late as the mid-eighteenth century.

Even though the 'Miner's Friend' found widespread use, especially in England, other engineers and inventors spent their careers trying to find new ways to utilise the full potential of steam. One of the most successful innovators, was an ironmonger from Devon named Thomas Newcomen. Born in 1663, by his early thirties Newcomen had considerable knowledge of metals and mining and believed he could improve upon Savery's machine. Around 1705 he went into partnership with a plumber named John Calley and together they built a giant steam-driven device which, unlike Savery's pump, used a cylinder and piston. Steam came from a giant beehive-shaped structure housed near the pump building, the steam was drawn off and passed into the cylinder and the top of this was sealed with a layer of cold water.

This first Newcomen pump was, however, remarkably inefficient and produced only a fraction of the power of Savery's machine. Furthermore, it was prone to leaking because the poor-quality solder used for the joints melted under heat. Two or three years of experimenting with this system found Newcomen and Calley making little headway until a fortuitous accident brought Newcomen to a startling conclusion, which he recorded in his journal.

The cold water that had been allowed to flow into the leaden jacket surrounding the cylinder penetrated the cylinder wall through a casting fault mended with tin-solder, which had been melted by the steam. Forcing itself into the cylinder, the cold water immediately condensed the steam, causing such a high vacuum that the weight hooked onto the end of the little beam, which was supposed to represent the weight of the water in the pumps, was insufficient, and the air pressed with such tremendous force on the piston its chain broke and the piston itself knocked the bottom of the cylinder out and smashed the lid of the small boiler. The hot water flowing everywhere convinced also our senses, that we had discovered an incomparably powerful force.[5]

Newcomen was not slow to appreciate the potential of this discovery and he soon constructed a system to recreate the conditions which had led to the accident for each stroke of the piston. A jet of cold water was sprayed into the cylinder just as the steam had built up sufficiently, moving the piston up and down. Connected to a sturdy beam, the piston was then used to evacuate water from the mine and also to pump air into the shaft using a bellows.

Newcomen's device was not very efficient but it nevertheless became a huge commercial success and it revitalised the mining industry. Larger and more powerful Newcomen machines were built to drive deeper and deeper shafts and a large proportion of the mined metal was used by the military for the manufacture of cannon and guns. But, ironically, it was the weapon makers themselves who facilitated the next leap in steam technology and helped produce an efficient steam engine that would find wide application beyond the need for pumps.

The story of how James Watt invented the stream engine after watching a kettle boil is, of course, like Newton's apple or Einstein's train, another neat, oversimplified version of the truth and it says little about the slow, deliberate progress of science and engineering ideas that culminated in Watt's revolutionary work.

In 1763, some fifty years after Newcomen had installed his first machine in a mine in Staffordshire, James Watt, then a twenty-seven-year-old instrument maker and general maintenance man at the University of Glasgow, was asked one day to repair a demonstration model of Newcomen's machine. He had taken little notice of it before but as he dismantled the model he was immediately struck by how inefficient the device was and how slowly it worked. He quickly concluded that if the steam was condensed in a chamber separate from the piston and cylinder it would work better. After constructing a test model he went on to build the first full-size and truly efficient steam engine.

This story suggests that posterity has been unduly kind to Watt. Although remembered for his great contributions to the harnessing of steam, Newcomen is not as famous as Watt, a name familiar to every schoolchild. This is because Watt took a very limited and largely inefficient device and turned it into something extraordinary – an engine that produced significant power, enough to move a train, enough to launch an industrial revolution. He succeeded in doing this because of two improvements he made. The first was his introduction of the separate condensation chamber. His second, and equally important innovation, was to make sure the piston fitted tightly into the cylinder, an improvement that greatly increased the power of the system.

Both of these modifications had come from a consideration of Newcomen's device. The first could be accomplished by reorganising the components, thus effecting a new design. The second, though, was dependent upon the ability of craftsmen to make tightly fitting components, and for this Watt was fortunate enough to have associations with weapon makers who were working on ways to produce better machine parts.

During the first half of the eighteenth century engineers worked hard on trying to improve cannon design but little real progress was made. A Swiss engineer and gunfounder named Jean Maritz was the first to reconsider the way cannons were made. Traditionally, the cannon was cast from iron (or, more usually, bronze in England) and it was shaped by a mould. However, this process created a less than perfect barrel because the moulds themselves were not made with the degree of accuracy required. Maritz was convinced that this failing explained most of the disastrous explosions of cannon.

Rather than casting his guns, he instead started with a solid cylinder of iron and, using a boring machine, he opened up a barrel. Between 1720 and 1740 he experimented with different designs, and his son (also named Jean) continued his father's work into the 1750s, but neither man could build a reliable and accurate boring machine that produced a cannon that worked better than the moulded type.

Then, in 1773, the French government established a commission to investigate why so many guns were exploding on the battlefield and to find ways of improving their design to increase safety. A French brigadier named Marchant de la Houlière decided to take up the challenge and he revived Maritz's scheme. De la Houlière concluded that the fault with his predecessor's method lay with the boring apparatus and so he developed a very accurate cutting head lined up with a guide bar that could bore a straight and smooth barrel through the cylinder of iron.

The year Marchant de la Houlière created this method he took his work to England and settled in the town of Coalbrookdale in Shropshire, near the iron foundry that had first put the town on the map. There he met two engineers, John and William Wilkinson, who happened to know James Watt. De la Houlière's cannon boring machine was precisely what James Watt needed to manufacture pistons and cylinders that would fit together snugly so that he could utilise the full power of the steam in his engines.

Three years after Marchant de la Houlière's work, Watt's first steam engine was put into service in a coal mine. A few years later, in 1782, he built the first rotative engine which produced far more power. By the turn of the eighteenth century, more than five hundred rotative steam engines were at work in mills and mines throughout England. The modern technological age had arrived.

Castles and Cannon

As cannons became safer to use, cheaper to produce and therefore more commonplace, the design of the fort or castle, against which they would be used, had to keep pace. The problem was that castles were sitting targets. Throughout the five hundred years between the building of the first European castles and the time of Leonardo and Machiavelli, the castle had grown steadily grander in appearance. The castle of the early fifteenth century had towering walls to defend it against slings, arrows and siege weapons. The walls were crenellated to protect defending archers and they were usually constructed from large blocks of stone.

During the early days of the cannon, castle design changed little and if a castle was redesigned at all it was usually to increase the height of the walls, which was believed to provide better defence. But in reality such a modification made it more of a target for enemy artillery. By the 1520s, new techniques in metallurgy and better quality gunpowder made cannon far more accurate and efficient, destroying forever the notion that a castle could remain impregnable. This in turn led to a complete change in the way forts and castles were designed.

The simplest way to protect a castle against cannon was to cover the towers and the walls in soil, which could absorb some of the energy of the projectile, but this was not an entirely satisfactory means of defence. It was ugly, expensive and time-consuming and so architects were forced to make fundamental design changes. First, the crenellations went and the height of the walls was reduced. Next, the walls were sloped outwards at the base to increase the degree of ricochet and thus reduce the damage a projectile might cause. Later, new castles were constructed from brick rather than trimmed stone as bricks could better absorb the energy of a cannonball. Even then, a cannonball from one of the new guns could bring down a tower or breach a castle wall, in the process killing scores of defenders. It soon became clear that if the castle was to survive at all it would have to undergo an extensive makeover.

The first move towards a genuinely different design strategy was the construction of a defensive barrier outside the main walls of the castle, and a deep ditch was dug immediately behind it. This system added a line of defence but it also meant that if the assailants broke through the outer wall the rubble fell into the ditch, blocking the attack, and allowed the cannon of the defenders to wreak havoc.

However, the truly radical change in castle design that restored the balance of power between attacker and besieged came once more from the application of ancient ideas that had been lost before the Dark Ages. These concepts were derived from Greek and Roman texts translated by Italian scholars during the late fifteenth century and they enabled architects and engineers to rethink completely the way fortifications were constructed.

With the application of these ancient ideas, the new castles that were built during the first half of the sixteenth century bore only a passing resemblance to the high-walled edifices of the pre-artillery age. These new fortress were often sunk into the ground behind an outer wall which meant that the attackers could only cause damage in the castle itself by firing over the walls. But most important was the change made to the design of the castle towers.

A well-known fault in the design of the traditional castle was the dead space which existed immediately beyond the walls and towers. This was an area in which attackers could shelter unseen from the battlements and it was almost impossible for defenders to reach them. From this position, assailants could tunnel under the castle and either launch a surprise attack once inside or plant explosives under the fortress. It was also a place in which a cannon could be positioned to be fired over the walls with relative impunity.

The new design involved the building of pointed towers or bastions that projected out into the ditch that surrounded the castle. These towers were elongated and had sloping walls so that a defender standing atop a bastion commanded a clear view into what would have been a dead space. Cannon positioned at the top of the bastions could fire back towards the castle ramparts, leaving nowhere for the attackers to hide.

This concept was developed gradually over a period of several decades and one of the most spectacular examples of such a castle is the Fortezza da Basso in Florence, designed by the architect Antonio da Sangallo and completed in 1533. It consisted of a complex arrangement of bastions that allowed defenders to see every point of the outer wall. Contemporaries considered the fortress impregnable.

Although such innovations often resulted in spectacular castles and succeeded in thwarting an attacker armed with cannon, they were slow to catch on because it was both labour intensive and expensive to remodel an old fort or to construct an entirely new one. Leon Battista Alberti (a man far ahead of his time) had written in 1440, almost a century before the completion of the Fortezza da Basso: 'If you were to look into the expeditions that have been undertaken, you would probably find that most of the victories were gained by the art and skill of the architects rather than by the conduct or fortune of the generals, and that the enemy was more often overcome and conquered by the architect's wit without the captain's arms than by the captain's arms without the wit of the architect.'[1]

Indeed, the architect's wit (if we extend Alberti's claim so as to include surveyors, designers, builders and theoretical mathematicians) was greatly enhanced by the arms race between cannon and castle. According to one sixteenth-century observer, Sir William Segar: 'Who without learning can conceive the ordering and disposing of men in marching, encamping, or fighting without arithmetic?'[2] This was entirely accurate because mathematicians had always played a significant role in military evolution, and, at the same time, they had gained reciprocally and channelled ideas back into civilian life. With the advent of the cannon, this process was exaggerated enormously.

The correct use of a cannon was no simple task. Cannon and ammunition were expensive and difficult to transport, so gunners needed to produce results. In the early days, trial and error was the best method for striking an enemy site, but soon the mathematician was being called upon in order to increase both accuracy and strike rate.

When the cannon first appeared, the standard instruments for measuring angle and height had changed little since the earliest days of this discipline. The astrolabe employed by the Egyptians and perfected by the Arabs of the ninth century was primarily an astronomical instrument used to measure the angle of elevation of the sun, planets and stars. Heights of buildings could be measured at a distance by a cross-staff, a device introduced into Europe by Levi ben Gerson who had translated Arab texts containing a plethora of long-forgotten mathematical secrets.

However, the greatest problem for the gunner was knowing the range, the precise distance from the cannon to the target. This problem was solved during the 1530s by a Dutch astronomer named Gemma Frisius, a professor of mathematics at Louvain University in Belgium. He took the simple step of putting the astrolabe on its side and creating the technique of triangulation.

To find the distance to the target, a gunner's assistant set up Frisius's device a known distance to the west of the cannon and measured the angle of the target from a predetermined north. Then he repeated the process to the east of the gun. Once the angles and the length of the base line (azimuth) had been calculated, a triangle could be drawn with the two long sides intersecting at the target. Dropping a line down from there to the base, the distance in a straight line from the gun to the target could be calculated with ease. The man who put all these procedures into one handy instrument was Leonard Digges who, in 1571, produced the theodolite, an instrument which is still employed by surveyors today. Triangulation is a technique also used universally in tracking and monitoring. It is the mathematical system at the heart of the GPS (global positioning system), modern communication networks and surveillance systems, where the gunner's assistant has been replaced by two satellites in orbit a known distance apart.

In the days when the only projectile weapons were siege engines, spears and bows, there was little need for a siege-engine operator or an archer to analyse trajectories; their work was governed by instinct and experience. Because cannons could be fired from much greater distances, careful analysis was required to ensure cannonballs reached their targets; consequently, with the invention of the cannon came the new science of ballistics. Leonardo da Vinci conducted a series of experiments to study the way in which cannonballs travelled through the air and the forces that influenced their motion. His findings helped him in his work as chief military engineer for Cesare Borgia, and many of his diagrams have survived in his notebooks.

Later, during the early seventeenth century, Galileo spent some time as consultant for the Venice Arsenal, devoting a great deal of thought and energy to devising a theory to explain how a cannonball flies. This led him to an early form of the calculus perfected independently by both Isaac Newton and Gottfried Leibniz a generation later. Leonard Euler, Francis Bacon and John Wallis were also fascinated by the mathematics of ballistics and wrote extensively on the subject.

However, the most important mathematical treatise to come from a study of ballistics is *New Principles of Gunnery*, written by the English mathematician and engineer Benjamin Robins and first published in 1742, a generation after Newton and Leibniz. *New Principles of Gunnery* became the standard work on the subject, a book written from first-hand experience of cannon performance and a deep appreciation of mathematical principles. Two hundred years after its publication, German rocket designers were still relying upon Robins's analyses to target London with their V1 and V2 rockets. During the 1960s, NASA engineers in the United States plotted a course to the moon employing the very same concepts and techniques applied on the battlefields of eighteenth-century Europe as quantified and documented by Robins.

Guns by the Million

A s the cannon evolved and countermeasures to defend against it became gradually more sophisticated, another form of ballistic weapon, the handgun, was beginning to find widespread use.

There is some debate about the earliest uses of the gun. It is believed to have been used first in China around 1300, and manuscripts offering drawings of the weapon in operation have been dated from that time. These, though, were little known outside China: only after the Arabs introduced the construction of the gun and the use of gunpowder into Europe during the early fourteenth century did the Western world first become aware of such things.

At first, there was great resistance to the gun; so much had been invested in the skill of the archer that the old-school commanders (who constituted most of the upper echelons of the army) were instinctively against change. It is striking to note that during a period of almost two and a half centuries between 1617 and 1850 the British Patent Office recorded just under three hundred patents directly linked to handgun technology, but in the single decade between 1850 and 1860, when gun technology improved dramatically, more than six hundred new patents pertaining to firearms were approved.

The earliest handguns were known in Europe as 'tubes' and they were indeed little more than pipes filled with a carefully measured quantity of gunpowder into which an arrow or stone was placed and then expelled by the force created by the gunpowder exploding. They were, of course, incredibly inaccurate and extremely dangerous to use. But it was clear from the beginning that, for all the early drawbacks, the handgun would prove to be an important and versatile weapon. Compared with an archer, a gunner required little skill or training to get impressive results; and, with the exception of a crossbow bolt, a bullet or an arrow fired from a gun was just about the only thing that could penetrate full armour. Even a relative amateur could, with luck on his side, successfully bring down a knight at thirty paces.

Such a weapon added significantly to the demise of the knight and a chivalric tradition already severely damaged by the reintroduction of the crossbow during the Middle Ages. Most importantly, the gun was relatively cheap and easy to make and its invention and use led to a radical change in the very structure of armies. Before the gun, power usually lay in the hands of a force with special equipment such as large cannon or a troop of heavily armed and armoured knights on horseback. The gun meant that the *number* of soldiers in a force became a crucial factor in the success or failure of an army. A large force of semi-trained peasants armed with handguns could readily overwhelm a body of the best-trained knights, pikesmen or skilled archers.

However, although creating a soldier from a peasant or a farmer by the simple expedient of giving him a gun was inexpensive compared with the cost of maintaining knights with their horses, armour and weapons, one of the disadvantages of the weapon was that (like a sword or a spear) it still had to be handmade by a craftsman. This meant that each gun was slightly different, each had its idiosyncrasies and each was, to a degree, unpredictable in its behaviour in the hands of a common artilleryman. It also meant that parts were not interchangeable, and, most importantly, it required a large team of craftsmen and artisans to supply the huge number of weapons needed by an army.

Advocates for a greater reliance on the handgun were aware of these shortcomings, and they knew they needed a method of producing large numbers of guns cheaply and quickly. Furthermore, they understood that guns should be made uniformly. In other words, the gun needed to be standardised. If standardisation could be achieved it would mean that there would be fewer unwelcome surprises in store for the soldier (many of whom already viewed the gun with suspicion) and training in the use of the weapon would be easier and more efficient. Lastly, if the gun was standardised then parts could be interchanged and damaged components replaced relatively easily; all of which made good economic sense to often cash-strapped rulers. In short, the gun needed to be mass-produced.

Ancient civilisations practised a form of mass production. Around 1400 BC Egyptian chariot makers pioneered the use of the spoked wheel which was far lighter and stronger than the traditional solid wheel. They were assembled in large numbers by a team of men; each member of the team was responsible for a single stage of the production process. In China around 200 BC, crossbows were also made in their tens of thousands very quickly using a form of production line in which each worker was responsible for a small part of the complex construction process. In ancient Greece, statues were mass-produced for export using a crude form of what is known as the pantograph principle in which a cutter connected to a tracer by a series of gears follows the contours of an original and duplicates it.

Once more, however, such skills seem to have been lost at the onset of the Dark Ages and they only began to resurface during the Renaissance when ancient texts were translated by Italian scholars. Cannon makers in sixteenth-century France were the first to introduce into their 'factories' some of the forgotten Greek and Chinese techniques. But, although the principle of standardisation was grasped quickly, for many years it was not developed beyond the idea that certain cannon should use mass-produced ammunition.

The adoption of a sophisticated factory system or a form of

mass-production that could churn out hundreds of thousands of guns did not take place because the technology required to replicate an item at a reasonable speed was simply not available. To make machines that could perform such tasks required specialised cutting tools, lathes, reliable gears and precision instruments. This meant that the advent of the mechanised production line had to wait until technology caught up and theory could be put into practice by innovative engineers and designers from Europe and later the United States.

The first major step towards a factory production line came about thanks to the innovative thinking of an engineer named Henry Maudslay when he was working at the Woolwich Royal Arsenal in south-east London. In 1800, he developed a new form of lathe, based upon the conventional model but scaled-up to an industrial-sized machine he called the 'go-cart'.

Around the time he built his first large lathe, Maudslay became friends with a royalist French émigré named Marc Isambard Brunel who, after travelling through America, had finally settled in London. Brunel had acquired a lucrative contract from the Royal Navy to produce ships' blocks – shaped blocks that form an integral part of the pulley system controlling the sails. In 1800, a ship of the line required anything up to two thousand such blocks, which, up till then, had been made meticulously by hand. As a consequence, it was often the delivery of these blocks that slowed down the building of a ship and delayed its launch.

Brunel immediately saw the value in Maudslay's lathe and realised that the blocks could be fashioned far more quickly using his device. It took the pair six years to develop the equipment, but, by 1808, their Portsmouth block-making yard had become the first large-scale mass-production unit, or 'factory', in the world where a staff of ten unskilled men could manufacture 130,000 ships' blocks per year.

This great step forward, coming just as the Industrial Revolution was blossoming, could have propelled industry in England even further ahead of her rivals but for the fact that this innovation was sidestepped and shunned by the Establishment. The reason for this was that the automation of what had been seen as a traditional craft was perceived as a dangerous threat. Many thousands of skilled craftsmen and the unskilled labourers who supported them relied upon the survival of the old trades, and the authorities feared serious social disorder if these men were replaced by machines. It took another war and clear evidence that the country was falling behind its industrial rivals to change English thinking on the matter.

As Maudslay and Brunel were developing the machines that would eventually fill commercial factories and help to make cars, bikes, sewing machines and a thousand other household commodities, the armaments industry in the United States was steaming ahead with methods of modernisation.

It was a Frenchman named Honoré Le Blanc who was behind this progression. He had developed a method of standardising parts for side arms by using a specially designed machine to make the components of a gun. He tried to interest the French government in his scheme but he was ignored. However, he had a more positive response when, in Paris, Le Blanc made the acquaintance of the US ambassador Thomas Jefferson, who quickly grasped the potential of the Frenchman's ideas.

America had been independent for a generation but her armed forces still relied entirely upon supplies from European manufacturers. Jefferson, as well as other politicians and military men, correctly identified this as a serious weakness. Soon after meeting Le Blanc, Jefferson returned to America and made it his business to convince Congress that they should invest research and development money in the Frenchman's plans.

For all his acumen and business sense and the support of someone as influential as Jefferson, however, Le Blanc was rooted in Europe and could not be persuaded to cross the Atlantic. And so the contract for making muskets for the US Army employing Le Blanc's methods was passed on to a well-known American inventor named Eli Whitney. Whitney adopted the invention of the production line for making guns as his own, and today he is recognised by most historians as the father of the factory system.

Yet he made almost no changes to Le Blanc's system, and he was not even terribly efficient at what he did do – he delivered his first 12,000 muskets to the US Army six years late.

Crucially, though (Whitney's lacklustre business performance notwithstanding), the idea of mass production very quickly captured the imagination of American entrepreneurs. In 1804, when Whitney finally delivered his first muskets to the army, America was still a young country driven by an adventurous spirit. It suffered from none of the social restrictions placed upon the English or most other Europeans, and even though many thousands of immigrants were arriving in the country each year, there was a chronic shortage of workers. An automated system to manufacture guns or anything else was therefore the perfect solution. It is not surprising that the innovation that rapidly transformed nineteenth-century industry was soon being referred to as the 'American System of Manufacture'.

Meanwhile, in England and France little changed even as British industry, built upon manpower and steam, boomed and its influence spread around the world. Factories employed vast numbers of unskilled men, women and children; and although the output from these engines of the Empire made England competitive, rich and powerful, the innovative spirit that had sparked the Industrial Revolution in England a generation or two earlier inspired no new manufacturing methods.

At the Great Exhibition of 1851 the cream of British industry, along with fifty thousand exhibits from thirty other nations, was put on display. It was an amazing spectacle. Forty thousand visitors a day descended upon the newly constructed Crystal Palace in London's Hyde Park, and during the six months of the exhibition some six million people attended. As a commentator of the day remarked: 'All social distinctions were for the moment merged in the general feeling of pride and admiration at the wondrous result of science and labour exhibited in the Palace of Glass. Never before in England had there been so free and general a mixture of classes as under that roof.'[1]

One of the exhibitors in Hyde Park in 1851 was the American

gun manufacturer Colt, which greatly impressed the visiting military delegations as well as the English gun makers who visited their stand. Even so, the British government remained complacent about modernising army weapons and it was not until a closely fought war shook their confidence that there came a change in attitude.

The military establishment claimed that the 1856 Anglo-French victory in the Crimea was in large part due to the efficiency of their long-distance arms supply system but this was not strictly true. As the press of the day eagerly revealed, supply lines were not maintained quite as efficiently as the military would have liked to believe. In particular, the slowness with which damaged or destroyed weapons were being replaced and supplies of ammunition reached the front line almost lost the war for Britain and France.

This close-run thing, at a time when the British Empire was at its peak, shocked and angered military commanders. By the time of the Crimean War it was also clear how radically the successors of men like Whitney had transformed industry in the United States. By the mid-nineteenth century America had overtaken Great Britain as the world's most successful industrial nation. American products were much cheaper than their British equivalents, across the industrialised spectrum. During the previous fifty years the Americans had made two great leaps forward: they had successfully adopted modern techniques to supply their military and they had adapted those techniques for general commercial use, mass-producing innumerable household goods.

Nevertheless, before the Crimean War was over the men who had visited the Great Exhibition and had narrowly missed humiliation on the battlefield had set in motion the establishment of an English equivalent to Colt. This worthy competitor was the Enfield Company, founded in 1855. Using manufacturing equipment imported from the United States, the Enfield Company began to produce guns in 1859.

Now the march of progress was unstoppable. The old manufacturing methods were rapidly superseded in Britain and throughout Europe, and their model was the 'American System of Manufacture', begun by a Frenchman but brought to commercial fruition by the particular genius and uncluttered social structure of American business.

The most famous example of how civilian mass production came from wartime needs is that of Henry Ford's car plants of the 1920s, which will be considered in Part 4. However, this technique, first used to supply armies, was also in large part responsible for two outstanding achievements of civil engineering during the twentieth century: the Panama Canal, the pet project of Teddy Roosevelt and one of the great wonders of the industrial world, completed in 1914; and the Empire State Building, its 103 floors constructed in the heart of New York City in just 410 days, between March 1930 and May 1931.

Both of these projects required huge resources and employed tens of thousands of men. Each involved the intricate meshing of planning and labour and each relied on the supply at the right time of thousands of components. A large number of these components were identical (including pillars, struts, walls, floor plates and buttresses), all produced in factories far from the construction site, but each component had to fit perfectly into the jigsaw puzzle of the construction. Without the principle of mass production behind such plans, they could not have been conceived.

Explosives

So far in Part 2 I have only considered one way in which guns, cannon and castles – the 'hardware' of the machines of destruction and the defensive shields – have influenced technology. But of equal importance is the role of 'software', the explosive.

The Chinese of about the ninth century are generally credited with the invention of gunpowder. *The Essentials of the Military Arts* (Wujing zongyao), published in 1043 and written by a number of authors including Zeng Gongliang, Ding Du and others at the order of Emperor Renzong (a ruler during the early Song Dynasty), mentions an older text from about AD 850 in which gunpowder is described. However, it was not until the publication of *The Essentials of the Military Arts* itself that a standardised formula, along with possible military applications, appears.

The ancient Chinese were extraordinarily warlike. The different tribes and ethnic groups within China were constantly feuding even while they were under threat from external forces, most particularly from the less advanced but avaricious Mongols on their western border. They were also very interested in the uses of fire and *Military Arts* contains illustrations of the most primitive incendiary devices imaginable – oxen and boar sent to charge the enemy lines with flaming spears strapped to their bodies and their tails set alight. Later, military designers came up with carts filled with straw and flammable materials which were set on fire and propelled into the enemy camp.

At the time *Military Arts* was written the gun had not yet been invented. The earliest examples of a device resembling a gun, a bamboo tube used to propel arrows using gunpowder, came at least a generation later and there is no hint as to its inventor. Indeed, from the evidence that has survived there is even some doubt concerning how much gunpowder the Chinese used in the theatre of war. For such a belligerent people it is surprising that they did not take the idea as far and as quickly as they could, but then it may simply be that all records of the earliest Chinese guns have been lost.

The Chinese were certainly keen innovators and although their culture so stifled the individual that it was virtually impossible to change one's social status, the government was wise enough to reward innovations, particularly those in weaponry. According to an eleventh-century document: 'In the third year of the K'ai Pao period of the reign of Sung T'ai-Tse [AD 969] the general Feng Chi-Sheng, together with some other officers, suggested a new model of fire arrow. The Emperor had it tested and when the test proved a success he presented the inventors with gowns and silk.'[1]

Europe was introduced to gunpowder by Arab alchemists some-time during the early fourteenth century, although how the chemical knowledge of making it reached Italy first and then the rest of Europe almost immediately after remains uncertain. Gunpowder was probably one of those inventions that was stumbled upon independently by different innovators. The Chinese are credited with being the first to record a formula for gunpowder, but before their knowledge reached other parts of the world European alchemists and natural philosophers had almost certainly come up with their own blend.

The most significant of these European proto-scientists was the great English philosopher Roger Bacon. A Franciscan monk, Bacon was both a devout priest and a radical thinker. He wrote many books in which he documented his fascination with natural philosophy, alchemy and the occult. However, such musings eventually led to his downfall. At the age of sixty-three, on his way to Rome to present to the pope his latest work, three far-sighted tracts, *Opus Majus*, *Opus Minor* and *Opus Tertium*, he was held by Church authorities on the charge that his books contained 'suspected novelties'. He lived out all but the last year of his life in a Vatican cell.

The work in which Bacon describes gunpowder is *Epistolae de Secretis Operibus Artis et Naturae et Nullitate Magiae* (*Book of Secret Operations and Natural Magic*). In this mid-thirteenth-century text, Bacon offered a workable formula for gunpowder (41.2 per cent saltpetre, 29.4 per cent charcoal and 29.4 per cent sulphur) and he may well have produced the mixture and conducted experiments with it. However, Bacon's problem was that, as a priest, he could hardly make his research public because in his time such investigations were considered occult and therefore heretical.

Ironically, within just a few years of Bacon's private experiments, gunpowder was used for the first time in Europe, during the siege of Seville in Moorish Spain in 1247. By the end of the century, there were references to gunpowder factories in Germany. A few years later, a widely read book, written by a character known as Mark the Greek, appeared in which the formulas for gunpowder of different varieties were detailed.

Gunpowder quickly reached a point of near perfection. That is, it was simple to blend, easily manufactured and extremely powerful, a method of killing not easily improved upon. Different chemists experimented with alternative combinations of the three essential ingredients – potassium nitrate (saltpetre), sulphur and charcoal – and they devised ways of overcoming the greatest drawback of gunpowder, its hydrophilic nature (in other words the ease with which it absorbed moisture). The material used to propel bullets and cannonballs in 1750 was almost identical to that employed in 1500.

By the eighteenth century chemical theory was becoming better understood, making gunpowder more potent and safer to use. In his role as commissioner of the Royal Gunpowder and Saltpetre Administration at the Paris Arsenal, immediately before the French Revolution, the great French chemist Antoine Lavoisier applied his considerable intellect to improving the quality of gunpowder and its method of manufacture. During a nineteen-year tenure at the Arsenal he succeeded in both increasing the purity of the constituents and improving the granulation process.

During this time, Lavoisier more than doubled the production of gunpowder, so that by 1787 the French Army had stockpiled five million pounds of explosive and Lavoisier's tinkering with the formula had increased its power, thereby increasing the range of French cannon from 150 to 260 metres. Indeed, this so worried the Royal Navy that they rushed to find a chemist comparable to Lavoisier to help formulate a better blend of gunpowder. When the French Revolution began to escalate and the British stopped the French acquiring many of the imports they needed, including supplies of saltpetre, Lavoisier succeeded in synthesising it using nitric acid and potassium salts. Unfortunately for Lavoisier (and for science), in 1794 he found himself on the wrong side of the ever-shifting political divide in revolutionary France and he was executed as a traitor. At his hearing the prosecutor showed astonishing ignorance of even the most elementary ways in which the great Lavoisier had helped the revolutionary cause, declaring '. . . the revolution does not need scientists'.

For near on half a millennium, gunpowder was the only reliable explosive available, and apart from its becoming easier to use, faster to make and slightly more effective, it changed little. Then, in the middle of the nineteenth century, a Swedish engineer and chemist named Alfred Nobel developed the first radically different explosive since Bacon's day.

Ironically, Nobel was a pacifist and was once quoted as saying: 'Perhaps my factories will put an end to war even sooner than your congresses: on the day that two army corps can annihilate each other in a second, all civilised nations will surely recoil with horror and disband their troops.'[2] When he was confronted with the potential of a substance called nitroglycerine he intended simply to develop it as an explosive that could replace gunpowder in open-cast mining. Until Nobel, the work of miners, road builders and demolition workers was made especially dangerous by gunpowder's temperamental nature. Nobel recognised that nitroglycerine, synthesised by treating glycerol with a mixture of nitric and sulphuric acid, was susceptible to vibration rather than heat, and invented a combination of nitroglycerine soaked into a porous substance he called dynamite. He patented this as: 'An Improved Explosive Compound: Dynamite' in 1867 (Patent Number 78,317).

The military were, of course, very quick to grasp the value of Nobel's new material and other inventors began to use it to change the design of guns and cannon. Meanwhile, Nobel himself continued to pursue research into new, more peaceful applications of his invention, but because he owned the patent for dynamite, military applications brought him financial rewards. He died in 1896 with an estimated personal fortune of the equivalent of some £10 million, part of which went towards establishing the annual Nobel Prizes in 1901, Nobel's way of trying to compensate for the fact that he had made so much money by creating a deadly weapon.

Since Nobel, many new forms of explosive, including ammonium nitrate, which is also used in a different form as a fertiliser, and TNT (trinitrotoluene), which is many times more powerful than dynamite, have been synthesised. Yet the most important aspect of this evolution is that the chemical knowledge gained from the constant need to find better, more powerful explosives has not only altered considerably the face of warfare, it has given rise to the very objects we use every day.

Soon after Lavoisier's execution in Paris, his assistant at the Arsenal, a young chemist named Eleuthère Irénée Du Pont, fled Europe with his wife and two children. They became part of the great influx of immigrants into New York at the beginning of the nineteenth century and within a few months they had moved south of the city to settle in the Brandywine Valley between Washington and Philadelphia.

Du Pont had brought with him the skills and knowledge he had acquired under the tutelage of the great Lavoisier and was an expert in the production of high-quality gunpowder. The timing of his arrival in the United States was propitious. The country was labouring under a trade embargo with Britain and France and desperately needed innovative minds like Du Pont's – a man who could fashion materiel from locally available raw materials. Furthermore, this was an era during which America, not so long ago a British colony, was on the cusp of expansion; the continent was ready for exploitation by industrialised society and the architects of this expansion understood that lack of materials could slow progress.

Du Pont's earliest efforts were to supply gunpowder to the mining and building concerns that were spreading throughout the eastern seaboard, but there was also a constant demand from the military. In excess of two hundred thousand miles of road were laid in America during the eighty years between Du Pont's arrival and the 1880s, but also during that time the armament requirements of the War of 1812, the Mexican War of 1846–8 and, most importantly, the Civil War of 1861–5, turned the DuPont Company (as it became known) into a vast organisation employing some of the most accomplished chemists in America.

The two world wars of the twentieth century earned DuPont billions of dollars from the sale of explosives but by the beginning of the century the company had already started to diversify its interests. Immediately after the end of the First World War it became a major shareholder in a fledgling concern known as General Motors, and in the following decade DuPont also invested heavily in the production of synthetic dyes. This venture failed, however, because DuPont could not compete with the established European dye manufacturers. But the company benefited enormously from this diversion because it established one of the most advanced laboratories in the world, which attracted many fine scientists. Indeed, at one point during the 1920s, DuPont employed more than one in five of all the chemistry Ph.D.s in the United States.

With this foundation and a desire to recoup some of the losses (an estimated $40 million) they had suffered in the dyes market, DuPont's management and shareholders were understandably keen to see some results from their incredible research and development facilities, and, although it was slow in coming, when it did the accomplishments of the DuPont labs would make history.

In 1928 the company employed one of the most brilliant chemists of the day, Wallace Carothers, who had already become a star in the Harvard Chemistry Faculty. Seven years later, on 28 February 1935, he and his team invented nylon. Trials and tests took up another two years of research under the most secretive conditions, a mindset already familiar to employees of the company because of its military connections. By Christmas 1937, the first synthetic ladies' stockings (or 'nylons', as they were quickly nicknamed) were in the shops.

The invention of nylons and other plastics owed a debt to the work on explosives conducted by DuPont and other chemical engineering companies around the world. Perhaps the most important link between these two inventions was the discovery in the early 1870s that nitrocellulose, a chemical closely related to Nobel's nitroglycerine, could, with the correct combination of solvents, form a compound known as a polymer that was strong, malleable and relatively inexpensive to produce. DuPont produced the first polymer from nitrocellulose and the firm marketed it as Celluloid.

To theoretical chemists this breakthrough hinted at a vast range of potential polymers, each synthesised by combining simple molecules with suitable linking agents to form long chains. Each chain exhibited a variety of properties and the synthesis could be adjusted to produce the desired effect. If a hard, strong material was required, by combining the right chemicals a chemist could synthesise a compound that would show such characteristics. If a thin but strong fibre was needed, a different set of reagents and conditions was called for.

It sounds easy. But, of course, a great deal of intelligent guesswork, high-powered theory and trial and error went into (and still goes into) forming the enormous range of plastics available today.

Nylon was the first multipurpose, mass-produced plastic and it evolved out of all the factors mentioned above. DuPont supplied the money and the facilities, and their innovative and brilliant chemists supplied the brains; during the eight years it took to develop nylon, the tremendous knowledge acquired by trial and error from the chemical tradition handed down by Lavoisier supplied the other requirements.

By the beginning of the Second World War the word 'plastic' had entered the language. As well as providing the material backdrop for the postwar world, during the six-year conflict nylon was the key component in four million parachutes, half a million aircraft tyres and countless numbers of ropes, ties, flak jackets and filters used in everything from jeep engines to blood plasma analysers. This new material proved particularly critical in compensating for the wartime shortage of metals, especially aluminium and brass.

Today it is difficult to imagine a world without plastic. It is astonishing to remember that this amazingly versatile collection of compounds has become so much a part of our everyday lives thanks to a long line of researchers, beginning with an anonymous Chinese innovator and continuing through the efforts of countless chemists working in laboratories all over the world.

The Bomb

The bomb is another weapon whose origin, it is believed, can be traced back to ancient China. The ancient Chinese fascination with fire led military engineers to devise incendiary devices long before they thought of propelling projectiles using an explosive mixture. Their simple incendiaries were constructed from flammable materials to produce what was really an elementary form of grenade. These devices were thrown at the enemy by hand or using siege engines. For the purposes of our story, however, we should turn our attention to one bomb-making endeavour that has changed the world more than any other: the Manhattan Project.

The ideas that inspired the making of the first atomic bomb have their roots in a group of mainly European theoretical scientists working between 1915 and 1940. The science they were investigating, evolving out of the nineteenth-century classical ideas of the atom, was a new field called quantum theory. The leaders in this new discipline were the German physicist Werner Heisenberg, the Danish physicist Niels Bohr, the Austrian physicist Edwin Schrödinger, the Frenchman Louis de Broglie, the American Robert Oppenheimer and a small cadre of other physicists working on the cutting edge of theoretical science.

Albert Einstein was a colleague of these men but he did not

support many of the more extreme ideas of the quantum pioneers. At the heart of their investigations was a radical new picture of reality. The new discoveries of quantum theory forced physicists to consider all things as probabilities rather than certainties, and, most importantly for the foundations of nuclear technology, showed that energy and matter are interconvertible.

As quantum theory became more and more sophisticated during the 1920s and 1930s some of the central premises of this new way of thinking seemed, to the outsider (and to some insiders), increasingly bizarre and, at times, counterintuitive. But this embryonic science could also explain the true nature of the atom and quantify the awesome power locked up within the subatomic realm. Ironically, although Einstein fundamentally opposed many of the tenets of quantum theory as it stood during the 1930s, it was an equation – the iconic $E=mc^2$, which had been part of his own theory of relativity – that allowed the quantum theorists to reveal just how powerful the atom could be.

What this simple equation says is that, via a process called nuclear fission, a tiny amount of matter may be turned into a very large amount of energy. The key to this enormous power is the letter 'c' in the equation. This modest, lower case letter stands for the speed of light and it has a value close to 300,000,000 kilometres per second. Square this number and you arrive at the figure 90,000,000,000,000,000 (9×10^{10} billion or ninety million billion). Using this equation it is easy to calculate that just one kilogram (2.2 pounds) of suitable material (an unstable radioactive isotope) can produce some 25 billion kilowatt hours of energy. This is equivalent to the energy needed to power twenty-five billion average household heaters for an hour. In an atomic bomb, this titanic amount of energy is released in a microsecond.

The scientists working in quantum science during the 1930s were aware that, in theory, the power of the atom could be harnessed to destructive ends but at first none of them believed that it was possible to construct an atomic bomb; the gap between theoretical understanding of the science and the technological expertise needed to unleash atomic energy seemed too great.

However, in January 1939, just as the Second World War was about to break out in Europe and the Nazis were unfolding their plans for world domination, a paper written by two German radio-chemists, Otto Hahn and Fritz Strassman, and the Austrian physicist Lise Meitner appeared in the journal *Naturwissenschaften*. In this paper, the authors showed via the results of a series of small-scale laboratory experiments that it was possible for an unstable isotope, when bombarded with neutrons, to undergo fission, releasing a large amount of energy.

This result stunned the global physics community, but at first little was done about it. It was one thing to produce a tiny effect in a lab, quite another to utilise it in the making of a workable bomb, and most scientists saw it as little more than an amazing but impractical discovery that might one day lead to fascinating theoretical lessons.

But not everyone was so short-sighted about the potential of such a device. Many of the cutting-edge physicists working in England and America during the late 1930s were émigrés from Germany, most of them Jews who loathed the Nazis. Although they remained on congenial terms with most of the scientists who had stayed in mainland Europe (the non-Jewish scientists led by the German physicist Werner Heisenberg), some émigrés were afraid that the German scientists would succumb to pressure from the Nazis and collaborate with them to attempt the seemingly impossible: the construction of an atomic bomb.

One such émigré scientist living in the United States in 1939 was the Hungarian physicist Leo Szilard. He was a young, middle-ranking scientist who had secured a college teaching position, but although he had little influence himself he did have powerful and important friends. Szilard was the first to appreciate the danger presented by the findings of Hahn, Strassman and Meitner. In his opinion, it was a distinct possibility that Werner Heisenberg, who was generally regarded as being amongst the top three physicists in the world, would already be working on a way to scale up the fission process and to utilise the power of the atom in a practical bomb.

Yet even with his influential friends behind him, Szilard's warnings to the government and the military went unheeded in the United States. Perhaps if he had been residing in England, a country on the very brink of war and much closer to the German threat, his voice would have been heard, but in America no one in political power was interested in the doom-laden portents of stateless scientists. Szilard understood this and concluded that to make his views heard he needed to enlist the support of more important figures within the scientific establishment.

Through the Italian Nobel laureate physicist Enrico Fermi, who had moved to the United States with his wife, Laura, in 1938, Szilard succeeded in arranging a meeting with Albert Einstein who had been working at the Institute of Advanced Study in Princeton, New Jersey, since 1933. Szilard was a good talker and passionate; Einstein detested Hitler as much as any émigré, and he was perfectly aware of Heisenberg's brilliance. After two more meetings with Szilard and Fermi, in July 1939, a matter of weeks before Great Britain declared war on Germany, Einstein agreed to write a letter to American President Franklin D. Roosevelt warning of the dangers posed by the latest findings in atomic science and urging the United States government to build an atomic bomb before the Germans beat them to it. The letter read:

> Sir,
> Some recent work by E. Fermi and L. Szilard, which has been communicated to me in manuscript, leads me to expect that the element uranium may be turned into a new and important source of energy in the immediate future. Certain aspects of the situation seem to call for watchfulness and, if necessary, quick action on the part of the administration. I believe therefore that it is my duty to bring to your attention the following facts and recommendations.
> In the course of the last four months it has been made possible – through the work of Joliot in France as well as Fermi and Szilard in America – to set up nuclear chain reactions in a large mass of uranium, by which vast amounts of power and

large quantities of new radium-type elements would be generated. Now it appears almost certain that this could be achieved in the immediate future.

The new phenomenon would lead to the construction of bombs and it is conceivable – though much less certain – that extremely powerful bombs of a new type may thus be constructed.

In view of this situation you may think it desirable to have some contact maintained between the administration and the group of physicists working on the chain reaction in America.[1]

Remarkably, even this warning from Einstein was ignored initially. The Americans were in a confused state over the war. They feared that Britain might be defeated, leaving them vulnerable, but public opinion was not yet on the side of becoming directly involved in the fighting. Furthermore, they did not trust the scientists at the centre of this latest scare. US Military Intelligence kept files on all the émigrés, and according to archives from the time, Szilard and Fermi were believed to be Nazi sympathisers and even Einstein himself was under suspicion. In the dossiers relating to Szilard, Fermi and Einstein, the officer consigned to the surveillance of the men, Lt. Colonel Constant, recommended that none of the three should be allowed to do any form of secret war work for the government.

Eventually, though, through a complex series of events the Americans were pushed into financing the Manhattan Project. One of the crucial factors was the personal involvement of Winston Churchill, who convinced Roosevelt of the danger of an atomic bomb. Second were reports that when the Germans invaded Belgium in 1940 they acquired a large supply of uranium (in the form of an ore mined in the Belgian Congo). The third factor was intelligence reports from Germany which suggested that Heisenberg was in fact making great strides in the creation of a German atomic bomb.

The Americans were slow out of the blocks, but once they were committed they threw everything at it. Officially begun on 6 December 1941, the day before the Japanese attacked Pearl Harbor, the Manhattan Project rapidly grew into the most expensive and ambitious undertaking in history: indeed, some have compared it to the building of the pyramids of Egypt.

The head of the project was General Leslie Groves, a conservative disciplinarian who never really understood the mindset of the scientists who worked for him. The chief scientist (who forged an often fraught relationship with Groves) was Robert Oppenheimer, one of the most brilliant physicists of his generation, a man who had worked closely with the pioneers of the quantum theory and who, by 1939, had made his own major contributions to this burgeoning science.

The idea of the Manhattan Project was a revolutionary one and it may be considered the first example of 'institutionalised war-science'. A special team of the best physicists in the free world was brought together under a military umbrella, a master plan financed entirely by the government with the sole purpose of building an atomic bomb before the Germans.

So vast and so all-consuming was the Manhattan Project that it can only be described in superlatives. During the three and a half years it took from inception to the dropping of the first atomic bomb on Hiroshima in August 1945 the project consumed $2 billion of government funding (equivalent to about $50 billion today) and the centre of the research, at Los Alamos in the New Mexico desert, employed thousands of scientists, engineers and technicians.

Amongst the team were some of the most respected scientists in the world, including Edward Teller, Hans Bethe, James Chadwick (the discoverer of the neutron) and a very young Richard Feynman (a future Nobel prize winner and iconic figure within the physics community). In 1943, they were joined by Niels Bohr who had escaped from Denmark; but most conspicuously absent were Fermi, Szilard and Einstein.

Even though at one time or another he suspected them all of being closet Nazis, homosexuals and deviants of some complexion, General Groves respected the mental prowess of the Los Alamos team, once referring to them as 'the biggest collection of egg-heads ever'. And there can be no denying that the development of the atomic bomb was an amazing feat of science. It was the first 'Big Science' project ever undertaken and it provided the template for later large-scale undertakings such as the Apollo missions, the Hubble telescope, the International Space Station, the superconducting supercollider and President Reagan's Star Wars efforts of the 1980s. What distinguishes these concerns from most other scientific endeavours is not just their size and the billions of tax dollars that go into them, but also the fact that they each combine a blend of the most advanced theoretical ideas with practical, applied science.

The making of the first atomic weapon relied upon theorists (who, we must remember, had only just learned that nuclear fission was a possibility) expanding upon the quantum science of some three decades and building upon a tight and cohesive mathematical understanding of the nature and properties of the atom. At the same time, the experimental physicists (working with the theorists) had an equally demanding task. They had to construct a practical system to purify the uranium isotope for use in the fission process and develop a suitable design for the bomb, the delivery system, the trigger mechanism and a complex array of other key components.

The success of the Los Alamos team was due to a combination of many factors. First, they were a group of brilliant men and women who were given the best facilities and every resource available at the time. But, just as important is the fact that they were also driven, and the politicians were driven by the fear of losing the race to beat the Germans. The very idea that Hitler could obtain an atomic weapon was too horrible to contemplate.

It was only after the war was over that it was discovered that the German scientists never came close to building an atomic bomb. They were provided with only a fraction of the resources available to the team at Los Alamos; and it should also be remembered that, for all Heisenberg's brilliance and the talent of his small team, between 1942 and 1945 the majority of the world's top scientists were in the deserts of New Mexico, not in Berlin.

The impetus to achieve a nuclear weapons capacity before the Germans was an extremely powerful one and the belief that they were in a close race was held with genuine conviction by the scientists at Los Alamos and almost all the politicians involved. However, it is a sobering fact that towards the end of the war, and several months before the bomb was used against Japan, British intelligence reports confirmed that German scientists were at least three years away from creating a practical bomb.

It is almost certain that this information was known to only a tiny inner circle of politicians and military men and that even Oppenheimer was unaware of this fact during the final months leading up to Hiroshima and Nagasaki. How much General Groves knew of this matter is also obfuscated, but he was most likely privy to intelligence reports and knew that there was no great race on after all. In any event, by this point so much money, time and effort had been spent on the Manhattan Project that the Allies felt that it was an unstoppable freight train and they were compelled to press on.

For the American government the question of whether or not to stop the project was a complex one, not least because the President and his close associates had to consider the long-term repercussions of their decision. By the time it was discovered that the Germans did not present a real atomic threat Hitler was all but defeated and Japan was struggling to survive. The Americans, Russians and British were already eyeing the future, the new world order that would follow victory.

The Americans were the only true victors in the Second World War. They had made an unimaginable fortune from the hardships faced by their allies, countries that had held the line and shielded America between 1939 and late 1941, when the United States was finally pushed into the conflict by self-interest. Money and resources flowed from America to Europe like water from a burst pipe during the war, and, after it was all over, when the debts had been calculated, there remained only two important powers on earth – Russia (which had lost the most in human terms) and the United States. The Americans wanted nuclear weapons as a way of ensuring global domination after 1945. There was never any chance they would terminate the Manhattan Project even once they had learned there was no serious competition from Germany; and indeed, when the war was over, some within the American administration tried (with only partial success) to keep nuclear secrets from their closest ally, Britain, a country that had supplied many of the best minds responsible for the success of the Manhattan Project.

During the summer of 1945, as the bomb was nearing readiness, the option of invading Japan with conventional forces was considered, but strategists calculated that it might involve the loss of around 36,000 American servicemen – an unacceptable price. The atomic bombs dropped on Japan are believed to have killed an estimated three hundred thousand civilians almost immediately and perhaps a similar figure again from the effects of radiation, loss of infrastructure and the collapse of medical services in the country. Not surprisingly, the ethics of using the bomb remains a matter of debate, but what is undeniable are the far-reaching and wide-ranging benefits that have come from the development of this weapon.

Just as the success of the project depended upon a symbiotic relationship between theoretical and experimental science, the spin-offs from the Manhattan Project and the ongoing research that has been conducted since that time have fallen into both camps – a greatly enhanced understanding of theoretical physics and many practical applications.

The advances in theoretical physics that came from Los Alamos are by their very nature less obvious to the public than the technological developments, but in their own way they are every bit as important. Furthermore, the progress in understanding the theory of atomic science has itself led to an acceleration in the evolution of technology.

The Second World War and the decision to create the Manhattan Project came at an auspicious moment in the development of atomic theory. For some three decades the relatively small group of experts in this field had been working within a loose multinational network. They had gathered at conferences to discuss their latest ideas and had published papers on their findings in respected scientific journals. The coming together of a large number of these scientists under one roof at Los Alamos created an incomparable think tank. These men and women were living in the same village-like community for more than three years. All of them had the same objectives and goals. Each was driven firstly by a desire to learn, but there were also rivalries and personal agendas that added an additional spark to their working environment. Combined with these motivations was the impetus from the politicians, men who were spending vast amounts of public money to achieve something that, during the early 1940s, lay at the furthest reaches of science and technology.

From the Manhattan Project came the very notion of the 'Big Science' project now routinely funded by government, and among the major spin-offs are the state-funded, high-energy atomic research establishments such as those at CERN near Geneva, and Fermilab, a two-hour drive to the west of Chicago.

At these establishments scientists have been working for several decades trying to understand how sub-atomic particles interact. They investigate the energy produced in atomic interactions and study a whole range of new particles that have been discovered using giant colliders. This machinery accelerates sub-atomic particles close to the speed of light and smashes them together to create new particles in the debris. These 'particle accelerators' evolved directly from the machines built from scratch at Los Alamos between 1942 and 1945 and they are now used to test the highest-level theoretical constructs of modern-day physics.

Particle physicists often work with pen and paper, computer terminals and a lot of RAM, but they need to verify their ideas because a solid theory is just an ephemeral hypothesis until experimental evidence can be found to support it and results repeated, documented and fitted into the greater scheme of science. Cyclotrons and other such machines in the high-energy research labs do exactly this. If evidence can be found, they are used to find it. As the theorists develop more and more exotic theories and suggest the existence of ever-stranger particles and components of particles, the accelerators are pushed further and further to quantify the concepts. This process often requires more and more powerful machines and correspondingly greater funding.

There is a constant battle between the scientists and the fund-providing politicians who often do not understand the relevance of cutting-edge theoretical physics, and one major project (the superconducting supercollider) has been cancelled in recent years because the costs of the experiments (running into the tens of billions of dollars) has been deemed too high.

When we consider the technological dividends of the atomic weapons development programme we naturally think first of the initialisation and growth of the nuclear energy industry. By 2004, 335 reactors in thirty-one countries supplied 16 per cent of the world's energy needs. But this is only one spin-off. The engineering and design challenges faced by the Los Alamos team required a very steep learning curve and the application of novel, lateral thinking. All of this meant that the generation of physicists, chemists, engineers and military designers which followed the men who worked on the Manhattan Project was given a tremendous boost by the achievements of three and a half years of intense, no-limits resourcing and the gathering of brilliance that characterised the task.

There have been some false starts in applying nuclear technology to civilian life. Perhaps the most notorious was Operation Ploughshare, a project to use controlled nuclear explosions in mining. Some nuclear technology advocates have been vociferous in their insistence that nuclear explosions can be employed to remodel land features and to carry out deep excavations. The man sometimes dubbed the 'father of the hydrogen bomb' and one of Oppenheimer's key collaborators at Los Alamos, Edward Teller, was evangelical about the merits of nuclear technology in macro-mining and geological remodelling. During the 1950s, small-scale experiments were carried out in Alaska as a trial for a much larger project. The plan involved using six huge hydrogen bombs totalling 2.4

megatons (40 per cent of the total firepower expended during the entire Second World War) to blast the north-west coast of Alaska in order to create a keyhole-shaped harbour near Point Hope. Residual radiation from the experiments remains to this day.

Other spin-offs have been far more practical and beneficial. In 1949, four years after the war was won, the United States government established Sandia National Laboratories. With several centres of operation close to the site of the Manhattan Project, Sandia is managed today by Lockheed Martin (having been under the stewardship of AT&T between 1949 and 1993). The primary responsibility of the organisation is to manage and maintain the nuclear arsenal of the United States, but in fulfilling this task a huge and constantly growing list of innovations has come from the Sandia labs. The remit for the organisation is a broad one.

As well as their primary task, they also monitor the proliferation of nuclear weapons, develop access codes for the top-secret establishments where the weapons are kept and model military scenarios. Sandia scientists have created advanced monitoring and surveillance systems using remote sensors both on satellites and within terrestrial platforms. They have developed the art of encryption and code construction, and, perhaps most dramatically, they have been responsible for the entire technology of computer modelling.

This last achievement, which during the 1990s spawned entirely new industries, began in the meeting rooms and cafeterias of Los Alamos between 1942 and 1945. The team, we must recall, were working with merely a set of possibilities. No one had made an atomic bomb before. No one knew its potential, no one could say with any certainty whether or not atomic theory could be translated into hard reality. With the exception of a single experimental device exploded at the Trinity test site 120 miles south of Albuquerque a few weeks before the bomb was dropped on Hiroshima, the team could not build scaled-down models of their weapon. Furthermore, aside from a brief period during 1945 when the team had access to a very primitive computer called ENIAC, they could work only with pen, paper and slide rule. So, they created imaginary scenarios with diagrams and numbers. You could almost say that they did computer modelling before the invention of a modern computer.

Today, we marvel at the stunning graphics used in computer games, we are thrilled by the computer-generated images in Hollywood blockbusters and most of us use high-powered computers with dazzling screensavers and windows operating systems. Perhaps this would not have been possible without Oppenheimer and his colleagues brainstorming in dingy rooms in Los Alamos or the development of their working methods by the individuals who followed them.

Another vastly important spin-off from the nuclear weapons research programme, one that affects the lives of many people, is nuclear medicine. One aspect of this advance is the use of radiation therapy for the treatment of cancer, but there are many more elements to this important strand of medical science. Radioisotopes are used to study fractures and malfunctioning organs and are being employed as a trace to investigate blood and lymph disorders and digestive tract problems.

At the core of the Manhattan Project was the effort to purify the material used in the first bomb, Uranium-235 (U-235), which had to be separated from the far more common isotope U-238. Because the U-235 isotope constitutes only 0.1 per cent of uranium, to obtain enough U-235 for the first bomb large quantities of unrefined uranium were required. A central separation plant built at Oak Ridge, Tennessee, employed more than half the total staff involved in the Manhattan Project and the task of separating useable quantities of U-235 took more than 50 per cent of the total budget of $2 billion. At the peak of its operation this single plant used more energy than the entire automobile industry of America and the government was forced to requisition silver bullion from the Treasury for the enormous electromagnets at the heart of the purification system.

The extractor used at Oak Ridge was a machine that could separate isotopes by capitalising on the fact that isotopes are each affected differently by a powerful magnetic field. From this work came the technology of magnetic resonance imaging (MRI), now employed in every major hospital in the industrialised world to create 3-D images of the body.

Some of the more recent ideas and innovations coming out of the Sandia labs sound like science fiction (it would be fair to say, anyway, that radioisotopes, supercomputers and smart materials would have been considered science fiction by the Los Alamos team that developed the bomb). Such cutting-edge developments include 'robugs' or nanomachines, insect-size machines that actually resemble bugs, that could be used in large numbers to carry out surveillance operations on the battlefield. The development of these devices has greatly accelerated the entire science of nanotechnology, which began to filter down to civilian research during the early 1990s.

Nanotechnology is expected to become one of the most important sciences of the twenty-first century and it will find applications in almost all aspects of society. Just two examples may illustrate this point: vast numbers of robot devices no larger than a single cell in your body and each controlled from a central computer will be used surgically to remove cancerous tumours from patients and to heal wounds and broken bones. Similar machines will be used to manufacture commodities 'from the bottom up', making an item from each tiny constituent part, in a precise imitation of Nature.

Nanotechnology is an embryonic discipline. Having been spawned by military scientists and planners, it has yet to make a true impact upon our lives. But soon it will follow mass production, the computer, the jet airliner, the nuclear power station and the antibiotic as a mainstay of our civilisation, a further example of the way a military agenda may paradoxically produce much that is good, elevating and inspiring.

PART 3

FROM

THE CUNEIFORM

TO THE

CREDIT CARD

Organised thought and the application of logic lie at the heart of any civilisation. Without these things it is impossible to rule, to manage a society or for any given culture to dominate others.

There are many forms of organised thought, but in this section I will consider four major forms and how they have influenced the advance of science and society.

The first is writing. Writing is something most of us take for granted. In a world deluged not just with images and sounds, but with words – billions, trillions of words – it is impossible for any literate person to imagine being unable to read and write, let alone to imagine a time when no written language existed. From its origins in ancient times, writing has been developed into a crucial tool in all advanced societies. Created originally as a system to record important events and to share and preserve ideas, reading and writing remains perhaps the most important skill we possess.

The second form of organisation is banking and the management of money. It has been said that bankers run our lives for us, and this may be shown to be as true for nations and corporations as it is for most individuals. Money is needed to create industries that in turn generate wealth. Money is needed to maintain the infrastructure of society, and money is needed to buy and to develop weapons. Impede the flow of money and a nation may be quickly destroyed.

The third consideration is logistics. A hidden thread in the structure of any military organisation, logistics offers little in the way of battlefield glamour, but it is essential to the maintainance of armed forces, and through military application logistics has been applied to aid the ordering of all modern societies.

The last factor is the management of time, the importance of which was only realised by military planners during the early nineteenth century. However, once its importance was noted, time-keeping became an essential ingredient in any military success, and a practice now at the heart of all organised societies.

The Written Word

Unlike the first demonstration of the telephone or the first flight of an aeroplane, no precise date can be placed on the emergence of the written word, but, as with many inventions, it derived from several sources. The Egyptians around 3000 BC, the Sumerians of Mesopotamia of about the same time, the Chinese of the fourteenth century BC, and, most recently, the Mexican Indians some 2600 years ago, each, in isolation, appear to have developed a form of writing.

The Sumerians' first writing is the one we understand best. Known as cuneiform, it uses pictograms and logograms, a lexicon within which each symbol represents a spoken word or concept. Cuneiform was an enormous cultural development but the form it took was also immensely clumsy and difficult to use. Each pictogram or logogram was a complex drawing, time-consuming to reproduce on stone tablets or, later, papyrus. In addition, to convey meaning and to be of greater use the lexicon was gradually expanded until it consisted of an ungainly four thousand symbols.

Even with these drawbacks, cuneiform was adopted as a common mode of communication and record-keeping that flourished for many centuries and spread beyond its place of origin. However, during the era in which cuneiform was used, very few people were literate and written language remained in the hands of a small elite of scribes and scholars who used it to record commercial transactions, religious ceremonies, royal events and proclamations, and, most crucially, to prepare documents for military strategists.

One of the oldest recorded cuneiform texts comes from the Hittite people of Bogazköy (modern Hattusas in Turkey). Here were found some 25,000 stone tablets comprising a huge resource of cuneiform records dating from the second millennium BC, perhaps five hundred years after writing was first invented in neighbouring Sumeria. One of the most important of these cuneiform documents is a record of the Treaty of Qadesh between the Hittites and the Egyptians. This well-known treaty of 'eternal peace' guaranteed security throughout the region for a generation. Another famous cuneiform record from the first millennium BC tells us of the exploits of the Assyrian ruler Nebuchadnezzar and his attempt to invade Egypt in 568 BC.

Limited as cuneiform was to a specially trained elite, the emergence of phonetic language therefore offered a huge leap forward. Again, it is difficult to date this but it seems clear that the Egyptians had an alphabet by about 600 BC and the Hebrews and Greeks by at least 400 BC. These advances were contemporaneous with the innovation of papyrus, which was superseded later by parchment.

These inventions had their source in the Middle East but the Chinese also developed writing independently. A form of symbol writing first appeared in China almost two thousand years after it was first devised by the Sumerians and Egyptians, but it advanced at a faster pace, and the Chinese were almost certainly the first to invent ink, writing implements similar to pens, and, later, paper. Ink and pens probably first appeared sometime around 1000 BC, and were used to draw pictograms and symbols on animal skins and wood. Paper, also credited to a Chinese philosopher, Lei-yang, who wrote about it during the first century AD, was not widely used for at least another thousand years.

The decisive event that brought paper and ink to the West was a conflict between the Arabs and the Chinese in the eighth century. This war culminated in the AD 751 Battle of Talas River in which

Arab forces led by Qarluq Yagbhu were victorious. Amongst the many thousands of prisoners taken at the battle was a large company of paper makers who were impelled to give up the secrets of their guild. From these experts, knowledge of paper making spread first to Samarkand and then to Baghdad. From there it was carried to Damascus, Cairo and Morocco, and finally through Spain and Italy to the rest of Europe, where it made a huge cultural impact.

With the replacement of stone tablets and chisels with papyrus, paper, ink and simple writing utensils, writing became 'mobile' and its military value grew rapidly. Before long, writing was being used for much more than just recording campaigns and glorious victories. It was quickly adopted as an important communications medium and a way to organise armies, draw up plans and supply lists and construct a chain of command.

The essence of military success lies with organisation. Guns and ammunition are important, numbers are important and such things as tactics and strategy equally so, but without organisation the best equipped and the fiercest armies can be crushed easily. The pen is indeed mightier than the sword and when the two work together successfully they are invincible. Military leaders throughout history have known this and it is some measure of the importance of writing to an army's success that we can discern a direct relationship between the size of a society's army and the level of its literacy.

During ancient times, when legions of scribes kept records of the fighting armies and maintained detailed accounts of weapons and other resources, numbers of men in uniform, costs and gains, generals such as Alexander the Great and Julius Caesar or the Ch'in warriors of fourth-century BC China were able to control forces of up to two hundred thousand men. By the sixth century, however, by which time literacy had all but vanished from Europe and the continent had been cut off from its supply of papyrus from the Middle East, armies never consisted of more than twenty thousand soldiers. Even these relatively small forces proved extremely difficult to manage and each army was only ever a short-lived gathering of men who were paid for the duration of a battle or a short campaign before being allowed to return home to the farm or the marketplace.

As Europe emerged from the Dark Ages, education improved and resources became more readily available, the size of armed forces grew correspondingly. Through the Renaissance, armies and navies became gradually more professional, more disciplined and far larger. By the early days of the First World War, twenty million men had been mobilised by the countries involved.

But the importance of writing to the advancement of nations was not limited to the ability to communicate across battle lines. As Jared Diamond has eloquently commented: 'Writing marched together with weapons, microbes, and centralised political organisation as a modern agent of conquest.'[1]

Writing allowed commanders to give explicit instructions to naval fleets searching for new lands and eager to beat down their rivals. Writing could be used to record accounts of previous missions and expeditions, and perhaps to describe the pitfalls, the dangers and the treasures to be found in new lands. Furthermore, writing was essential to both the organisers of such missions and those with the responsibility for running a colony once it was established. It allowed for the organisation of the economic and domestic life of the new dominion and it was used to manage imperial ambitions from the other side of the world. Furthermore, writing made sense of map making and maps made sense of exploration which led ultimately to practical colonial exploitation.

The invention of movable type in the West in the fifteenth century was a development that was abhorrent to the Church.* This was because it offered the potential to educate the masses and there was nothing the Church feared more than education. In some ways their anxieties were well founded for the invention of a cheaply reproducible written language made it possible for those with the necessary skills and material resources to communicate their ideas to a broad section of society.

---

*I add the caveat 'the West' here because movable type was actually first invented by Bi Sheng in China around AD 1050.

The mass media began with the Romans who from about 60 BC had a daily news sheet, the *Acta Diurna* ('Daily Events'), said to have been created originally by Julius Caesar. It was hand-written and freely available. Coupled with this was a newsletter covering the proceedings of the Roman Senate called *Acta Senatus* (literally 'Proceedings of the Senate'). A thousand years later, something similar to the *Acta Diurna* was popular in medieval Europe, but because very few people could read, its contents were more usually conveyed to the public by a town crier.

Later, the *Notizie Scritte*, first published in Venice in 1556, revived the idea of a daily paper and soon after this a quite different form of newsletter was created called the pamphlet. Such publications covered a wide range of subjects, from news stories to gardening advice, and were distributed widely across Europe. The pamphlet was also favoured by intellectuals who, before the advent of the academic journal, used this means of communication to proselytise new scientific discoveries, inventions and details of explorations to far-flung corners of the world. Pamphlets were also used to tell the public of the changing fortunes of wars being fought in distant countries. One best-selling example, *The Trew Encountre*, printed in England by Richard Fawkes in September 1513, reported on the Battle of Flodden Field which had taken place in Scotland a few weeks earlier. In this publication he offered eyewitness accounts of the battle together with a list of the English heroes of the day.

Later still, religious propagandists of the 1620s created the concept of the 'news book', a refinement of the pamphlet which became extremely popular in Europe and in the colonies. But the first newspapers as we would know them today appeared in England in the early 1640s during a brief spell of publishing freedom between the abolition of the court of law known as the Star Chamber and the establishment of the Commonwealth in 1649. Some of the earliest newspapers included the *Mercurius Politicus* which was edited by John Milton, the *Spectator*, which began in 1711, and the *Tatler* in 1709, the last two of which are still thriving today though in a rather different form as magazines.

The Times was first published in 1788 and quickly became the most important newspaper of its day. The significance of this newspaper and a number of others that followed in its wake (including the *Morning Post*, the *Observer*, the *Daily Telegraph* and the *New York Times*) was greatly enhanced by coverage of the Crimean War in the mid-1850s. William Howard Russell earned fame as one of the first war correspondents and his writings inspired Florence Nightingale to take up her mission to the Crimea. Although Russell declared that he was 'the miserable parent of a luckless tribe', within six years of the end of the Crimean War an estimated 150 war correspondents were working with the armies of both sides in the American Civil War sending back dispatches via telegraph and Pony Express to newly established media nerve centres in Atlanta, Washington and New York. From there, news of the war was communicated to London and the rest of Europe.

The reports published in Britain and France during the Crimean War did much to alert the public to the horrors of armed conflict. This was really the first time that the public had been given ready access to graphic and unbiased reports from the front line, and for many such news came as a terrible shock. This coverage did much to establish a new phenomenon – public opinion – which prompted questions in Parliament about the conduct of the war, and from then on public opinion played an important part in the political structure of Western nations.

The use of language, printing and the media could, of course, also be employed to the advantage of the government and the military, both in keeping the public quiet and in influencing the decisions of the enemy. As a psychological weapon propaganda has always played an influential role in the fortunes of nations and it is a tool almost as old as war itself.

Before people had the ability to spread their ideas through the written word via the printing press, they employed other means. Greek playwrights used the theatre both to put over often controversial political views and to explore differences of opinion on religious and social matters. Later, Roman writers delighted in defying the authorities by making use of drama to convey their political, social and moral opinions, views that were often thinly disguised so as to avoid imperial persecution.

By the Renaissance, as printing became popular propaganda was viewed as an essential military tool. In 1588, Philip II of Spain employed rather clumsy tactics of denial to hide the devastating loss of his armada, and, half a century later, both the Dutch and the English were accusing each other of military atrocities in far-off colonies via pamphlets widely circulated amongst the literate of the time.

The term 'propaganda' first came into popular use after the Church of Rome created the Congregation for the Propagation of the Faith, initiated in 1622 by Pope Gregory XV. This was a commission of cardinals charged with the task of spreading the faith and regulating Church affairs in 'heathen lands'. Gregory's successor, Urban VIII, created a College of Propaganda to train priests for these missions. But it was with the advent of inexpensive printing that propaganda really became an important factor in the campaigns of enemy governments as well as a weapon for revolutionaries and military commanders.

Anti-British propaganda was used enthusiastically by American colonies struggling for independence in the 1770s, and it was used again to great effect in the French Revolution when men such as Voltaire and Rousseau inflamed opposition to Bourbon rule in France. Using propaganda, Georges Jacques Danton and his fellow revolutionaries succeeded in maintaining public hatred for the monarchy and the deposed ruling elite, just as Tom Paine had steered opinion against the redcoats during the American Revolution.

Hitler was a master of propaganda, and thanks in large part to his personal efforts before the war began in Europe in 1939 the Nazis had established an influential propaganda department skilled in disseminating carefully contrived disinformation to their enemies. This was also employed ruthlessly to control the minds of the German people themselves. As Hitler's Minister for Propaganda and National Enlightenment, Joseph Goebbels masterminded one of the best organised and most influential propaganda machines in history.

'Why are we fighting?' asked a German newspaper in 1939 under instructions from Goebbels's office. 'Because we were forced into it by England and its Polish friends,' came the response. 'If the enemy had not begun the fight now, they would have within two or three years. England and France began the war in 1939 because they feared that in two or three years Germany would be militarily stronger and harder to defeat.'[2]

Six year later, in the dying days of the war, the propaganda machine was still pumping out ridiculous perversions of the truth. 'The defenders are the equal of the enemy,' one Berlin newspaper proclaimed as the city fell to Allied forces.

> Even with his superiority in men and matériel the defensive forces will come from troops in Berlin, from the great reservoir of the *Volkssturm* [military units consisting of those too young or too old for regular military service], and from the shortening of the front. The enormous number of buildings in Berlin is being daily added to by its citizens, according to the old military truism: 'It's better to sweat than to bleed.' Behind its ramparts Berlin is a military training ground. Its citizens are learning to use the *Panzerfaust* [an anti-tank weapon] and the machine gun. Berlin and its citizens are like a large army in an encampment facing a strong enemy.[3]

Both Germany and the Allies in the Second World War used the written word in an effort to influence the opinion of the civilian population of the enemy state. Bombers were risked in dropping tons of pamphlets over cities as well as over theatres of conflicts themselves to encourage desertion, and the relatively new medium of radio was employed by both the Allies and the Axis powers to try to turn public opinion.

Disinformation and propaganda are perennial instruments during any conflict and they are employed today just as readily as at any other time in history. Indeed, it might be argued that propaganda through pamphleteering, and more recently via the internet, is more effective today because a far larger proportion of the target audience are literate and therefore open to influence.

During the 1950s, the advertising industry began to boom. Before the war, the most noticeable forms of advertising were notices in newspapers and magazines, billboards and radio ads, but by combining the lessons learned from propaganda campaigns from the war and utilising the new medium of television, advertising gradually became woven into the fabric of all our lives.

From the 1960s, advertising grew rapidly into the pan-global business it is today; and it is impossible to deny that during the past half-century it has played a significant role in changing our society. An important element in the move towards globalisation has been what we might call 'accepted propaganda'. In other words, because many people are aware of the endless attempts by others to influence our attitudes and thinking on almost everything, from washing-up liquid to the morality of the Gulf Wars, we have the luxury of weighing up what the powerful corporations and the governments of the West have to offer. This is, of course, a by-product of technologically advanced democracies which allows citizens access to media and the ability to vote governments in and out of power combined with a legal system that, in theory at least, protects against corporate corruption.

Keeping Secrets

Almost as soon as a system of writing had been constructed, mankind found ways of taking it apart. This reaction was an impulse motivated by military need, for secrecy is elemental to the military mind and is a necessity during times of conflict. Encryption and decryption techniques became more sophisticated and ingenious with each generation and as they did so they grew increasingly important to the military establishment.

There are two ways in which codes may be used; these are called steganography and cryptography. Steganography is the physical concealment of a message, the most famous example of which comes from Herodotus, who wrote of a method used by the Persian Histiaeus who sent a message to Aristagoras, the tyrannical ruler of Miletus, by tattooing a message on the scalp of a slave and waiting for the hair to grow back. He then sent the slave to Aristagoras with instructions to shave the man's head.

Another ingenious variation on this idea was the skytale used by Greek commanders. This involved writing a message mixed in with random letters or words on a piece of papyrus. If this was wrapped around a stick, the message could be read along the length of the stick. The message was then sent without the stick.

The recipient would have to have the same size stick and know how to wrap the scroll in order to decode the message.

Cryptography, a far more versatile code system, has been favoured by military planners, commanders and governments since the earliest days of writing. Julius Caesar is said to have been one of the first soldiers in the field to use a code, sending messages from his campaigns in Britain back to Rome using the simplest of all cryptographs, shifting the letters of the alphabet by three, so that an A became a D, a B became an F, and so on. Only those who knew the shift could translate the code. This may appear remarkably unsophisticated today, but because it was one of the earliest codes to be used the sheer novelty of the encryption kept the secret safe, at least for a while.

During the Dark Ages in Europe codes fell out of favour along with reading and writing but Renaissance military men and philosophers rediscovered encryption. Leonardo da Vinci concealed his most secret researches by composing his notes using mirror writing. Roger Bacon was obsessed with codes and ciphers, and during the middle of the thirteenth century he wrote a widely read treatise on the subject entitled *Secret Works of Art and the Nullity of Magic*.

The polymath genius Alberti, who greatly influenced and inspired Leonardo in so many areas, has become known as the 'father of Western cryptography' because he introduced many of the key ideas still used by analysts today. These ingenious ideas include frequency analysis, a technique used to detect and define patterns in a text which then give important clues to the code key. He also made the first polyalphabetic ciphers and the earliest cipher wheel, using a series of wheels engraved with numbers and letters that could be used to substitute the letters in any given message.

Alberti's ideas for polyalphabetical coding systems were further developed by the German scholar Johannes Trithemius who published the *Polygraphiae* in 1518; Alberti's cipher wheels were adapted by Thomas Jefferson who used an elaborate set of twenty-six of them to create a code machine used from the early nineteenth century until it was retired by the American military in 1942.

Perhaps the most famous code story of modern times is that of the Enigma, an encryption device developed by the Germans before the start of the Second World War to encode the details of field operations and to communicate with their submarine fleet. The cracking of the Enigma code became a high-priority operation for the Allies and led to the establishment of a specialised team of British cryptographers and mathematicians at Bletchley Park in Buckinghamshire. They began decoding Enigma messages in April 1940 and continued to operate throughout the war. Their work not only saved thousands of Allied lives, but it greatly accelerated the development of the first computer. Most crucial was the construction of a machine called Colossus, a project led by Alan Turing and a small group of analysts who became the world's first computer specialists and paved the way for the massive expansion of computing that followed the war.* It is therefore not surprising that the development of the computer has been inextricably linked with codes ever since. Today, the lessons learned from cryptographers working for the military is of profound importance to business and science and it remains a valuable tool for strategists and politicians as well.

The most important application of cryptography has been the creation and development of data encryption, which is employed extensively in the banking sector and is a crucial component of e-commerce. We all take for granted the ease with which we can slip a credit card into the ATM and, using our personal code, withdraw money. Every day millions of people around the world buy goods on the internet and can do so safely only because of the use of the encrypted codes built into our credit cards.

Public Key Encryption is the term given to the system that allows us to use a personal security number or pin number to access information. This system was patented in 1975 by Whitfield

---

*See Part 7.

Diffie and Martin Hellman, a pair of American researchers at Stanford University. However, the method was actually devised two years earlier, in 1973, at the nerve centre of British Intelligence, GCHQ (Government Communications Headquarters) in Cheltenham. Here Clifford Cocks and Malcolm Williamson, two cryptographers working for the British government, produced the essentials of the method as a way of providing a flexible encryption system for managing Britain's nuclear arsenal. But, because they were constrained by the Official Secrets Act and their work kept highly classified, they could not make their findings public.

The Public Key Encryption system works on the principle that a long series of numbers, or a 'key', may be used to scramble sensitive information such as bank details or personal medical records. This scramble key is readily available to the public because it only encodes. Crucially, though, only someone with the decode key (unlocked by an ATM password, for example) can unscramble the original message or batches of information.

Mathematicians were the first to deconstruct language and to formulate codes using their understanding of numbers and patterns. However, such a process is a two-way street because mathematicians have also learned a great deal from the devising of codes and algorithms, and this knowledge may be applied in other mathematical disciplines as well as by workers in other fields. One extremely important example of this cross-fertilising is the way mathematicians collaborate with biologists and chemists to unravel the intricacies of one of Nature's great codes – the genome, the genetic imprint of all living things.

A favourite analogy employed by science writers to explain genetics is the written word. The entire genetic make-up of all living things is based upon DNA (deoxyribonucleic acid) which is comprised of just four molecules, known as bases. These are cytosine (C), thymine (T), guanine (G) and adenine (A). When discussing genetics, it is handy to talk about these bases being a little like letters. Groups of bases are then said to form the equivalent of 'words' and these 'words' may be arranged into 'paragraphs', an analogy for DNA molecules. These DNA strands arrange themselves to form a gene (analogous to a chapter of a book). Genes then group together to form chromosomes (books), which are packed into the nucleus of a cell, in much the same way as books are stored in libraries.

It is equally easy to think of this analogy in reverse, to visualise DNA as being like an encoded text so that the genetic code (as it is often called) determines where the bases go and how they inter-relate to form the DNA molecule. Further complex coding instructs the molecules of DNA to form genes and then chromo-somes. All of this means that scientists trying to understand better how genes work and how DNA does its incredible job have bene-fited enormously from thinking about them in much the same way as cryptographers think about cracking enemy codes.

Making the World Go Round

Money, in the form of token exchange and bartering, existed before writing. But in parallel with the development of ways of linking words together to create a written language, ancient cultures also devised ways of calculating, and, most crucially, recording numbers, which enabled a monetary system to come into being.

The importance of cash flow to military men became particularly apparent as innovation changed war with increasing effectiveness. Arming peasants with pikes and crude swords was a relatively inexpensive exercise when compared with the resources needed to keep artillery divisions armed and supplied with ammunition. It was also far easier to transport and control infantry than to move heavy cannon or siege engines. To wage war with potency, military leaders from about the thirteenth century onwards needed to have a strong financial base to work from or else they were obliged to borrow the funds needed to pay, feed and transport their troops, to supply ammunition and to keep a sizeable army (and perhaps a navy) fighting efficiently. The earliest banks were created to meet this need, to fuel the engines of war, and some of those who first realised the potential of financing military conflict became astonishingly rich and influential.

A form of banking was practised in ancient times; the Jewish moneylender, for example, was a figure portrayed in the Old Testament, and ancient records show that the earliest civilised peoples, the Sumerians and the Egyptians, dealt with money and understood the idea of lending and charging interest. But it was not until the thirteenth century that the earliest modern banks appeared in Italy, in the great trading centres of Genoa and Venice. From here the concept of an institution that controlled saving and lending and could regulate cash flow across the full spectrum of society spread to other regions of the country and later throughout Europe.

This change was facilitated by two innovations that had come to Europe from the East. The first was the concept of double-entry bookkeeping which allowed the maintenance of more accurate records of financial transactions and reduced the risk of expensive mistakes. The other was the invention of a Bill of Exchange, or cheque, which meant that business transactions could be made at a distance rather than solely at local branches of a bank. These two advances reached Europe just as scholars were taking possession of Greek and Latin texts that helped ignite the Renaissance, and they proved to be every bit as influential in moulding European life.

The most famous banking family in history was the Medici of Florence who made their fortune from supplying money to powerful leaders of Europe seeking to finance their military campaigns. Some historians consider the most prominent members of the Medici to have been dictators and point out that many of them were corrupt and motivated by self-interest. These claims are reasonable but there can also be no doubt that for some three centuries the Medici were *the* family of the Renaissance, the most important and influential citizens of Europe, or that they added enormously to the evolution of Western culture.

The first of the line to come to public prominence was Giovanni di Bicci de' Medici who in the early fourteenth century created a bank specialising in lending money to the Catholic Church. Giovanni was responsible for taking the Medici from middle-class respectability to prominence as one of the wealthiest families in Florence, and when he died he left an estimated 100,000 florins (equivalent to about £50 million, or $90 million today) to his descendants.

Cosimo de' Medici, Giovanni's eldest son, was born in 1389. Although he quickly demonstrated an acute business sense and a gift for numbers and ledgers, he was actually far more interested in books and the new learning arriving in Europe from the East. Between the ages of fourteen and seventeen, he was taught by a scholar of Greek named Roberto de Rossi who excited Cosimo's taste for ancient intellectual property just as Giovanni was preparing his son to take up the family baton at the bank.

Nevertheless, Cosimo became a superb banker, surpassing his father in almost every way. He led the family business skilfully, increased its wealth enormously and became the most respected and admired figure in Florence; but he also kept faith with his love for the arts. Throughout his life, Cosimo patronised artists, writers and musicians. He sponsored seats of learning and paid his servants to scour Europe and the East in search of ancient manuscripts, the translation and reproduction of which he also financed. In this way Europe's most powerful financier took money from warring leaders and spent a good proportion of it enriching the culture of his time as well as producing a legacy for the future.

Cosimo de' Medici also contributed enormously to the political and social stability of Florence by forging strong links with the financial institutions of other countries. During his lifetime, the Medici bank opened branches in France, Germany and England, all of which brought increased wealth and prosperity to the citizens of his beloved Florence.

Cosimo's son Piero succeeded him as head of the family and assumed the political role vacated by his father. Parsimonious and unimaginative, he was in almost every way the very opposite of his father. He outlived Cosimo by only five years and was succeeded in turn by perhaps the most famous Medici of them all, his son Lorenzo.

Although he was Piero de' Medici's immediate successor by genes and blood, Lorenzo was actually a throwback to the thrusting dynamism of his grandfather, in all ways Cosimo's true heir. He even showed his grandfather's initial reluctance to go into banking and politics, but once he did so he too proved a genius at making money. Although, like his forebears, Lorenzo was never officially the leader of Florence, the city and much of Italy were dominated by his presence.

Lorenzo was a gifted intellectual as well as an astute businessman but most of all he was a patriot who put Florence before all else. He demonstrated his devotion soon after taking over the bank. In 1471, a Franciscan monk named Francesco della Rovere, who came from a poor rural family, became pope. Taking the name Sixtus IV, he had been in office no more than a few months before he clashed with Lorenzo, the young banker, who, by controlling much of the flow of money in Europe, had the power to ground or to clip the wings of kings and pontiffs alike.

Sixtus believed he was destined to be more than a religious leader and moral guardian. He perceived himself as a leader of a temporal state, a warlord as well as a priest. He was also immensely greedy and avaricious, a man who possessed few moral scruples and considered the expansion of the papal states his first priority. Within months of ascending to the papal throne, Sixtus had set his sights upon either buying or taking by force large chunks of the north-eastern region of Italy known as the Romagna where he planned to install his nephew Girolamo Riaro as a puppet leader. But such a scheme cost money, and although in recent decades the papacy had become as rich as it had ever been (in no small part thanks to the guidance of the Medici bank) Sixtus was obliged to go to Lorenzo and ask for a loan of 40,000 ducats (the equivalent today of about £6 million, or $11 million dollars).

As far as Lorenzo and the Florentines were concerned, Sixtus's plans were unhealthy and would not only compromise their trade routes (which for centuries had run through the Romagna to Venice) but also threatened to destabilise Italy after years of peace acquired through careful democracy and statesmanship. When the new pope proposed his plan, it quickly became clear to all that he was motivated solely by self-interest and had little concern for the subtle balance of power in Italy. Lorenzo felt no compunction in turning Sixtus down.

In response, Sixtus felt equally justified in sacking his banker, which he did straight away. But this move did nothing to further his plans and he knew he had lost the first battle. Frustrated and furious, he failed to raise funds elsewhere for his scheme and so he retreated to lick his wounds. His simmering hatred for Lorenzo, whom he described as '. . . an evil, undutiful man who defies us', prompted him to declare publicly his wish to change the Florentine regime and have the Medici overthrown.

For six years friction between Rome and Florence chaffed diplomacy and the everyday course of Italian politics. Then, on 26 April 1478, four assassins hired by Sixtus's nephew Girolamo Riaro tried to kill Lorenzo. The attempt failed and the conspirators were caught and executed; the pope's power was diminished and Lorenzo acquired even greater popularity.

This series of events showed all the leaders of Europe where the real power of the times lay and it was not in the hands of hereditary sovereigns; men like Lorenzo could make or break entire nations. There was nothing any ruler could do about this. Power had shifted into the hands of the financiers and it would never be taken back completely. It could be said that war had created capitalism.

Even as banks became an intrinsic element of everyday life they continued to hold the purse strings of nations. The Bank of England was created in 1694 when William of Orange forged a financial partnership with an incredibly wealthy Scot named William Paterson. Paterson was a canny businessman who knew the King was in dire straights and needed an advance of over a million pounds to finance a series of campaigns against the French. Before agreeing to the loan he demanded that a national bank be established to manage the loan and that the government pay hefty interest. After a great deal of argument Paterson's terms were finally accepted and William was free to fight his wars.

The Bank of England was the first national bank in the world and it established the idea of the national debt. By 1700, the £1.2 million William had borrowed had grown to £12 million and little more than a century later, after victory in the Napoleonic Wars, it had ballooned to a staggering £850 million (approximately £100 billion, or $200 billion in modern terms). Indeed, the Bank of England had become so important to the government that in 1781 the Prime Minister, Lord North, declared in Parliament that it had '. . . from long habit and usage of many years become a part of the constitution'. He then went on to declare, '. . . all the money business of the Exchequer was done at the Bank, and as experiences had proved, with much greater advantage to the public, than when it had formerly been done at the Exchequer'.[1]

Another family with consummate power is the House of Rothschild which originated in eighteenth-century Germany before establishing a bank in London in the first decade of the nineteenth century. The bank grew to become one of the most powerful in the world by financing a succession of wars, including one of the most expensive (in terms of slaughter and hard cash), the American Civil War. One of the great figures of the family who established the Rothschild bank in England was Nathan Mayer Rothschild. He was said to have owned half the wealth of the world and reportedly crowed: 'I care not what puppet is placed upon the throne of England to rule the Empire on which the sun never sets. The man who controls Britain's money supply controls the British Empire, and I control the British money supply.' After hearing this, Thackeray quipped that Nathan Mayer Rothschild was 'Not king of the Jews, but the Jew of the kings.'

Conspiracy theorists suggest that the Rothschild dynasty, in league with other powerful financiers of the mid-nineteenth century, attempted to gain absolute political control through their financial activities and hatched a plan whereby the French and the British were to take opposite sides in the American Civil War. According to these theorists, the Tsar of Russia learned of the plot, realised the terrible danger inherent in the scheme and intervened to thwart the ambitions of the bank.

Such theories are probably fantasy and there is no hard evidence to support these claims, but in fact there is really no need for them anyway. The power of the Medici, their German contemporaries, the Fuggers, and others including the Rothschilds, the Rockefellers (USA) and the Barings (Britain), has been so great that these families have effectively controlled the fortunes of nations and the evolution of modern society more than any other institution including that of the British monarchy, the government, the armed forces or the Church. They have acquired this control (and continue to do so) by holding the keys to the bank vault and supervising the flow of money around the world.

The way banks manage cash flow during a conflict has done far more than secure the outcome of a battle or the chances of success in any given conflict. Banks have become fantastically rich, and with this wealth and concomitant power they have altered radically the way money is raised, managed and spent, not just for governments and big business but for almost everyone alive. They have also effectively changed the very structure of society.

Before the advent of banks, armies were raised from the local population. Peasants were made to fight with any weapon that was thrust into their reluctant hands. However, as weaponry improved and money became available through loans, this system changed. Leaders now wanted better troops with the skill to use the latest weapons. At the same time, civilians who had little appetite for war preferred to remain on their farms and in their offices rather than spill their own blood on the battlefield. So rulers raised money to pay soldiers through taxing the population. These taxes provided the leaders with the capital with which better armies could be raised and the people in turn were happier paying taxes than risking their lives.

The concept of taxation had existed since ancient times and it lay at the heart of feudalism, a system that had served most of Europe for at least a millennium after the fall of Rome. But the modern version of taxation, which has changed little over five hundred years, was inspired by the success of international banks and used to finance the technological upgrading of the world's armies.

At first, money raised by banks and from taxation led to the phenomenon of mercenary armies. However, the employment of such hired hands produced its own problems. Mercenaries were notoriously unreliable and if the money supply or resources were held up or fell short in any way they would desert their employer without hesitation. Equally dangerous was the fact that a mercenary army fought for whoever paid them, so they could easily be turned against their employer by the simple expedient of a better offer.

Before long it became clear that the best way to defend one's country was to create a professional standing army that was disciplined and controlled by elite officers drawn from the upper classes. This kind of army was expensive, but with a national bank in place and the facility of a national debt as well as a source of income from taxes, such costs could always be met. Almost without realising it, the hereditary leaders, and, later, those elected democratically, passed global power over to the world's bankers.

Organising and Logistics

The importance of banks to the military grew from the discipline of the banks' controllers. They knew how to handle money and how to monitor its whereabouts and its earnings. But any banker worth his salt also needs a ruthless streak and an all-consuming sense of self-importance. Indeed, viewed in this way, it is easy to see how the military worlds and the bankers coalesced, for those at the heart of both shared many qualities and characteristics.

In Part 3 I have considered two key areas in which military requirements have provided significant social change: 1) the military imperative, without which the evolution of language and the development of writing would have been much slower, and 2) the modern financial world, which stems from roots put down centuries ago by a union of military strategists and monetary experts.

A third way in which the lessons of war have shaped modern life is discipline, an aspect of the military which is today so taken for granted that it is viewed as almost synonymous with soldiery and the mindset of the fighting man.

The Roman army was a famously disciplined machine and this characteristic matched perfectly the society it protected. Class was an integral part of life in the Roman Empire and its class structure was clearly reflected by its military, which was also a many-tiered institution. Crucially though, civilian life in Rome benefited from important lessons learned from the need for discipline during military campaigns. Scribes, book-keepers, administrators and politicians often cut their teeth working within the strictly organised Roman army or navy. Many of the skills these men later brought to civilian life came from roles they had mastered in Gaul or Germania, Britain or North Africa.

As military technology grew more sophisticated, the rise of the specialist within the armed forces became much more pronounced. A marked change occurred when cannon and other firearms were introduced during the fourteenth century.

The gun and the cannon quickly became widely used weapons. Artillery troops were responsible for the maintenance of their weapons and for keeping themselves supplied with ammunition and gunpowder. These were tasks that required discipline and organisation and such men found they could apply these skills to a range of jobs when they returned to civilian life. This development in turn led employers to the belief that good soldiers often made good employees.

Today, discipline within the military is of greater importance than ever before. The technology of war has become so sophisticated and complex that a modern army employs a wide range of specialists. But, at the same time, the non-specialist soldier has little responsibility for hi-tech machines other than those he or she uses on a daily basis, such as communication devices and personal weapons, so, to operate most effectively they need to be disciplined to work as a team within a carefully controlled command structure.

In a broader sense, military prerogatives have led to the establishment of many institutions and systems used by government and big business to maintain a high level of organisation. In Europe, the civil service is a product of the military and is an institution that has been honed by discipline and organisational rigour over a period of perhaps four centuries.

The civil service operates with military precision and has an elaborate ranking system. It was created as little more than an off-shoot of the military, and to act as an intermediary between the army and the national government. Today, a strong relationship between the two institutions remains, and, indeed, many long-serving and high-ranking military personnel have exchanged commands in the army, navy or air force for positions in the civil service.

The tradition of a civilian institution that bridges government and the military is another legacy of the Roman Empire. The first-century AD philosopher and writer Pliny was one of the most important naval commanders in the Empire, but he was also a government servant, adviser, engineer and official historian. In more recent times, in seventeenth-century England, Samuel Pepys was both Secretary to the Admiralty Board and a high-ranking civil servant. Such men drew heavily on what they learned in the military arena and they applied this knowledge in civilian life.

Another way in which discipline has played a crucial role in maintaining armed forces is through the growing field of logistics, the science of planning complex operations. The value of logistics cannot be overestimated for, without intelligent planning, any military force can be defeated no matter how much firepower it musters. Indeed, logistics is perceived by most military thinkers as so important that it acts as a linchpin in any modern campaign.

While certainly true, this is not immediately obvious. Although many ancient societies invaded neighbouring territory, and both the Roman and Mongol Empires covered vast areas, their acquisitions were made over a long period of time and largely in a linear progression. In other words, each new territory was gained only after securing control of an adjacent region. During such campaigns an invading army would rely upon the surrounding countryside for supplies of food: it foraged, hunted or stole what it needed. But this was never a very reliable way of feeding troops and as armies grew much larger it became increasingly impractical.

The first commander to take note of the shortcomings of the traditional system of supply was Napoleon Bonaparte. In 1796 he invaded Italy with an army of 360,000 men, all of whom relied upon foraging to provide sustenance. During the Italian campaign, this system worked well because Napoleon split his army into small groups which then acquired supplies from a predesignated region. However, four years later, at the Battle of Marengo, his army was almost defeated by a far inferior Austrian force which attacked unexpectedly while much of the French force was away hunting for food.

Upon his return from Marengo, Napoleon, always a master of strategy, established what was called the Society for the Encouragement of Industry. His intention was to award prizes to those who produced the best ideas for improving the life of soldiers and for furthering industry. One of the great success stories to come out of this was that of a young inventor named Nicholas Appert who was awarded 100,000 francs for his remarkably useful innovation – food preservation.

Appert had grown up in a wine-making region and had been a cook and champagne bottler. He discovered that if food was placed in a champagne bottle which was then heated and sealed, the food would remain uncorrupted for long periods of time because the air had been driven out and so putrefaction was greatly slowed.

This was precisely what Napoleon's armies needed. Champagne bottles were cheap, the method of preservation easy to perform and cartloads of food could be transported over long distances with relative ease. Most important, without the need to forage, soldiers could concentrate on fighting. By 1812, French forces were marching on preserved food and within two decades Appert's champagne-bottled foods were on sale in Paris and London shops.

Although this method worked well and served both the military and domestic needs of the time, a decade after these preserved foods first appeared an English paper-mill worker and mould maker named Bryan Donkin succeeded in creating the world's first tin can as a stronger and lighter container for preserved food. Appert had considered the idea of using metal containers, but French industry was in such a bad state following the French Revolution that it had not been economically viable. Even in

England, where mining and metal forging were boom businesses, the earliest metal cans to reach the shops cost ten pennies (10d), about the same as the weekly rent of a family home.

The British armed forces enthusiastically accepted the new art of preserving food and storing it in tin cans. Cans could be stored aboard ship easily, they could be transported in carts and wagons, and, because the military required supplies in large quantities, the cost of this technology was greatly reduced. In 1818, the Royal Navy purchased 23,779 cans of meat and vegetables, effectively transforming the canning industry.

Ancient campaigns, and even those of Napoleon during the early nineteenth century, cannot, however, be compared with a modern war fought on several fronts simultaneously such as the Second World War. This was the first truly global conflict; no continent was left unaffected. Today, military historians point to one particular campaign from that war that marks the origin of modern logistical military planning far from home and conducted on an unprecedented scale. This was the second Battle of El Alamein in 1942, which ended in a famous victory for General Montgomery whose forces were pitted against those of the German commander Field Marshal Erwin Rommel, allied with a contingent of Italian troops and tanks.

The most important factor to determine the outcome of the battle was Montgomery's superior logistical planning which allowed him to build up a well-supplied and cohesive force of some million and a half men, tanks and artillery. This operation involved shipping vast quantities of food and ammunition, medical supplies, clothing, fuel and spare parts from England and other Allied countries to the battleground of North Africa. In the days before computers, the entire operation was planned with pen and paper and the carefully formulated instructions were followed to the letter by Montgomery's staff.

This victorious campaign in the desert established British supremacy in North Africa after 1942 and it may be seen as a rehearsal for the D-Day landings of June 1944, an operation that resulted in the Allied liberation of Europe. But beyond this, it is also the point in history when logistical analysis was acknowledged as an essential tool for any major military or peacetime operation. Careful structuring of supplies and manpower was employed at Los Alamos, on the beaches of Normandy in 1944, and later, in the space programmes of the United States and the Soviet Union during the 1950s and 1960s. The discipline required for such successful organisation also maintains the multinational corporations that today influence almost all aspects of commercial and political life.

Counting Heads and Watching the Clock

Ancient man conducted regular censuses. Part of the Romans' insistence upon social discipline, book-keeping and record-keeping is reflected in the importance they placed on regular monitoring of the population. For them the census was an essential tool for government overseeing of tax collection and land use. Two millennia later, during the eighteenth century, European and American governments constructed their own census systems based on the Roman model.

Ever since the Dark Ages civilians in Europe have always kept some form of birth, death and marriage records but it was not until 1597 that the Catholic Church made such record-keeping mandatory. Sixty-eight years later, in 1665, French Canada became the first modern society to carry out an official census, and in so doing it began a trend which was followed by the governments of other nations keen to monitor their own populations.

Part of the motivation behind this in French Canada, in Sweden almost a century later, in Italy in 1770, in the United States in 1790 and in Britain in 1801 was the need to collect and record figures for taxation and to monitor the ownership of property. But at the same time the census was important to the military, who needed accurate numbers of the population of men eligible for wartime service. By the nineteenth century, national service or conscription had become popular and provided a standing army of young men who were paid by their government to train and fight.

The reintroduction of the census (which is used today to provide far more than a simple head count) was of enormous benefit not only to the military but to government, to the medical profession, to sociologists and statisticians, to social planners, and for commerce, because it provided information about so many aspects of the civilian population. For the first time the distribution of people in towns, cities and rural areas and family structure and trends in birth rates could be analysed. Market research and the science of social categorisation grew from this and sociologists could quantify as never before the relevance and social impact of spending habits, the level of disposable income, the influence of education and even the diverse personal tastes of the population.

Another aspect of civilian life taken for granted today which had military origins is time-keeping. In the ancient world few people had any idea what year they were living in. Their day-to-day activities were governed by natural rhythms – the position of the sun, and at night by the movement of the stars and the moon. This was perfectly satisfactory for simple folk but as business became more sophisticated and international in character a reliable system of time-keeping became essential. It was driven by the needs of the military, but, even so, innovators had to struggle against the Establishment of the day to get their ideas accepted.

The great clock maker John Harrison produced for the Royal Navy remarkably accurate timepieces that enabled a seaman to measure longitude and assisted the navigator enormously, but his ideas were only adopted after he spent many years convincing staid and unimaginative commanders and politicians of the need to plan naval operations around precise time-keeping.

During the eighteenth century accurate clocks were large and expensive and consequently rare. Few outside the military could afford them or indeed had much use for them. Gradually, as timepieces became more sophisticated, smaller and cheaper, the wealthy elite began to place importance on punctuality and this attitude slowly spread throughout society. But even so, most military commanders had little interest in controlling campaigns with any attention to precision timing or synchronisation. During the American War of Independence, for example, George Washington, Commander-in-Chief of the Continental Army, never considered appending the time of the day or night to his letters and in his war correspondence there are almost no references to the time a battle or manoeuvre started or finished.

A generation later, in France, Napoleon Bonaparte had a very different attitude. It was his insistence upon careful time-keeping, punctuality and synchronisation of movements that brought these things into public life. He understood that considerable organisation, a keen awareness of time and the careful synchronising of regiments in different locations was of paramount importance in moving large armies across Europe. Even so, it took considerable persuasion on Napoleon's part to convince his generals to follow his line of thinking.

To be fair to those who had no appreciation of the importance of time, accurate time-keeping was, even as late as the mid-nineteenth century, a rather difficult exercise. The scarcity of good timepieces made it hard enough, but, to make matters worse, there was no consensus about time zones and each country and region operated within its own self-determined time-keeping system. Because the sun moves east to west through 15° of longitude every hour, before the problem of synchronisation was solved far-flung regions of a particular country had different ideas about what time it was. In England, for example, someone in Great Yarmouth on the east coast would think it was half an hour later than someone at Land's End in the far west. This disparity was quite unimportant unless military planners communicating by telegraph (as Napoleon did) needed to coordinate their operations. It also presented a problem for the compilers of railway timetables at the advent of the steam age in that trains often moved quite quickly between regions with different local times.

To circumvent this problem, a standard time longitude was established. The global reference point for this system became the Greenwich Observatory in south-east London, not far from the Woolwich Arsenal and Naval Academy, which led to the term Greenwich Mean Time (or GMT). From the 1840s this system was gradually adopted in Britain and across the Empire, and it became an international standard after the Washington Conference of October 1884.

Today, accurate time-keeping is an essential part of commerce, the financial markets and international communications. Indeed, twenty-first-century inhabitants of what we now call the global village could no more imagine an asynchronous world than sixteenth-century peasants could visualise living their lives to an agreed international standard or the concept of someone setting their timepieces to GMT.

PART 4

FROM

THE CHARIOT

TO THE

BULLET TRAIN

Until less than a century ago the fastest any human being had ever travelled was about 40 mph – the speed of the fastest galloping horse. The path that has taken human civilisation from this lack of technological advancement to a state in which people regularly speed along the freeway in their cars or travel at hundreds of miles per hour on bullet trains and in racing cars is not a straight one, nor is it a tale of development dominated by any single genius. Rather, the evolution of human mobility on land has relied on a matrix of ideas, inventions, discoveries and innovations.

The impetus for the creative ideas that spawned these advances has come from a number of sources. Some pioneers saw the commercial value of their ideas early, while others stumbled upon their contributions by accident or good fortune. Many innovations have come from a specific need, the knowledge that a piece of the jigsaw puzzle was missing.

As with so many of the innovations described in this book, military concerns have played a critical role in the development of our locomotion on land. Indeed, the two great wars of the twentieth century were particularly important in accelerating the improvement of sophisticated machines such as the car. This vehicle incorporates a large number of different components, from the pneumatic tyre on which it rides the road to the battery that sparks the fuel and powers its lights for night driving. Many of these components were developed for military use.

Man has always wanted to move faster, more comfortably and less expensively. The great minds of the Middle Ages and the Renaissance dreamed of an age in which machines could do the work of men and that what we today call technology could expand human horizons. Yet, for them, such things could never go beyond the designs they drew and the simple models they built. The infrastructure necessary to support such advances simply did not exist. Only after the Industrial Revolution, when gradual progression of invention and discovery provided the links between inventions and the technology to make horseless carriages and steam-driven trains possible, could the fantasies of the past be transformed into the reality of the present.

War has played a key role in providing these links. Soldiers need to travel to the battlefield, supplies and weapons have to be moved from one place to another. The innovation of weapons mounted on a moving platform provided an obvious advantage to any army. At the same time, vehicles must have roads to travel along, a fact appreciated long before the advent of powered vehicles; and with the road comes civilisation.

Horsepower

The earliest relationship between man and horse was that of hunter and hunted. Cro-Magnon man of about fifty thousand years ago viewed the horse in the same way he did any other animal, as food. Evidence for this hypothesis comes from archaeological sites in Salutre in southern France, where the remains of some ten thousand horses have been unearthed at the base of a cliff. Presumably they were driven to their deaths by prehistoric hunters and butchered to feed the tribe. By around 4000 BC the horse had taken on a different role, as a beast of burden. Nomadic peoples favoured it for this purpose because it was lighter and faster than an ox, it ate less and could carry almost as much as an ox. About this time nomads began also to ride their horses.

European and Asian nomadic tribes almost certainly understood the military value of the horse, for their lives revolved around a limited number of essentials – finding food, defending themselves and finding new lands and resources. Each of these essentials required, in varying degrees, aggressive behaviour and the horse provided a significant military advantage over warriors on foot. In the hands of the nomadic tribes of Central and Eastern Europe, the horse became a weapon in itself.

However, the horse's full potential was not realised until the beginning of the third millennium BC when the technologically more advanced peoples of the Near East domesticated the animal. Those who adopted the horse and integrated it into their society at this time had several advantages over those who had first utilised the horse. They had formed a civilisation rooted in agriculture and trade and they lived in towns and villages. Most importantly, they had learned how to use metals and to craft wood with some sophistication. It was this combination of adoption and innovation that led to a radical change in the military organisation of these cultures and with it came an equally radical change in the way they travelled and traded.

One of the first horse-drawn vehicles was the chariot. It was originally developed by a Near Eastern culture, the Assyrians, possibly around 1800 BC; but their design was greatly improved upon by the Egyptians some two centuries later. These vehicles used two spoked wheels, and, later, iron hubs (centrepieces) and greased axles. Within another two or three centuries, chariots were equipped with the equivalent of metal tyres made using an iron ring that was expanded by heat and dropped over the wheel rim before it cooled. As it cooled and contracted it formed a tight bond with the wooden rim.

Egyptian chariots came in two different models. The war chariot was lighter and more manoeuvrable. It had two six-spoked wheels and was usually pulled by two horses. The carriage chariot, a mainstay of traders and farmers, had two four-spoked wheels, was heavier and could carry greater loads; in effect, it was the precursor of the four-wheeled cart.

War chariots were expensive pieces of equipment but they quickly became essential to an army. The finest examples used by the Egyptian elite forces took up to 1600 man-hours to build. By about the twelfth century BC, the Egyptians and other advanced cultures of the region could boast chariot corps in which specially trained warriors using composite bows could cut a swath through enemy infantry and overwhelm less well-equipped forces. In skilled hands they were devastating weapons that could travel at speeds of up to 30 mph.

To the Greeks, who took the chariot to new levels of sophistication, the vehicle became as important to their military leaders as the tank would be to the generals and politicians of the mid-twentieth century. Homer waxed lyrical about the chariot as a devastating war machine, a crucial element of trade, and as a means of transport for the rich. In Book XV of *The Odyssey* one of his characters, Telemachus, stirs another, Pisistratus, with his heel, declaring: 'Wake up, Pisistratus, and yoke the horses to the chariot, for we must set off home.' In *The Iliad*, Homer also recounts how the commander, Diomedes, refuses to retreat before the overwhelming force of the enemy. Instead, he engages in a chariot battle and, guided by the goddess Athena, kills the warrior Pandarus.

As the peoples of the Near and Middle East were developing the chariot, the Chinese and Indians were developing sophisticated horsemanship skills independently. Three crucial innovations that were of enormous military benefit, and all of which appeared in China between about AD 400 and 700, were the horseshoe, the saddle and the stirrup.

Each of these inventions allowed warriors to travel further and faster, but for the military the most important was the stirrup which provided the means by which a rider could form a unique physical association with the horse, almost becoming an extension of the animal.

By about AD 600, the stirrup had been adopted with enthusiasm by European armies and within a century the structure of military forces had changed beyond recognition. According to the historian Lynn Townsend White: 'Few inventions have been so simple as the stirrup, but few have had so catalytic an influence on history. The requirements of the new mode of warfare which it made possible found expression in a new form of western European society dominated by an aristocracy of warriors endowed with land so that they might fight in a new and highly specialised way . . . all made possible by the stirrup.'[1]

The truth of this statement may be seen in the way European society changed during the seven hundred years between the earliest adoption of the stirrup in Europe and the fourteenth century. The earliest knights, elite warriors riding the largest and fastest horses, armed with spear, sword and axe, were 'shock troops' who levelled everything before them. In an age before the cannon or even the crossbow (which was used in Europe from about AD 1000) these riders were practically insuperable and their very presence shifted the balance of power. The military leaders who needed such all-powerful horsemen covered the enormous cost of maintaining such a force, but at the same time the men themselves became important and influential, eventually calling themselves 'noblemen'. The French origin of the name for such skilled horsemen, *chevalier*, gave its name to an ideal, the chivalric tradition. They were provided with land and the land was worked by a new class, the peasant.

This system was known as feudalism and it was the fundamental social construct of medieval Europe. It led to the evolution of powerful noble families who dominated European politics for centuries after feudalism was abolished, and its vestiges remain in the class system that continues to have significance and to influence society in the twenty-first century.

All Roads Lead to Rome

In prehistoric times, armies could fight only within their immediate surroundings. Even with the advent of the horse rider and the chariot, travel was severely restricted because the only routes anywhere across the vast stretches of the continents were barely passable natural paths or a rare few primitive, man-made tracks.

The Egyptians could travel the length of the Nile and through its tributaries by boat to wage war across stretches of desert. In the same era, the people of the barbarian lands to the north battled on the steppe and in the forests, but these conflicts always took place close to their homes and camps. The first conquerors to take war to far-flung parts of the known world and to sustain a vast empire were the Romans. They could do this thanks to the network of roads they built, along which they could transport troops, arms, food and supplies. Indeed, it is no exaggeration to say that the pickaxe used to build roads was no less important in constructing the Roman Empire than was the sword.

Rome was founded during the eighth century BC, but it was not until more than four centuries later, in 312 BC, that the construction of roads was first recorded. The most famous Roman road is the Appian Way, which runs south from the Porta Appia (today called the gateway of San Sebastiano). It was named after the

Imperial Censor, Appius Claudius. From about the early third century BC a network of roads grew up around the imperial capital and spread outwards from there to crisscross the Roman dominions. And although, in terms of both capital and labour, this road building was an enormously expensive project, it was seen, quite correctly, as essential to the expansionist plans of the Roman emperors. There could be no empire without a transportation network.

Today, the faint imprint of the Roman road system is still just about traceable in several parts of Europe. The main reason for the survival of this template is that the Romans built their roads with military foresight and vision and, in fact, their road plans have never been bettered. The roads were built as straight as was physically possible with the tools of the time and within the limits of geography, so that they took the traveller from A to B with maximum efficiency. As a consequence, all road-building schemes that imitated those of the Romans (including the European motorway system) followed the routes laid down during the days of the Roman Empire.

By the time the Roman war machine rumbled into Britain in AD 43, during the reign of the Emperor Claudius, imperial road builders had reached the peak of their profession. They began building roads immediately and progressed through the country as one region after another fell to their well-trained and heavily armed soldiers. With only a brief interruption caused by the revolt of Boadicea in AD 61, the Romans succeeded in dominating the entire British Isles apart from Scotland by about AD 77, by which time they had constructed a staggering eight thousand miles of road.

All the Roman roads were built by the army, a special corps of men several thousand strong, who followed the conquering troops as they moved from one village and town to the next along the existing Iron Age tracks. The fighting force sent to quell the tribes of Britain was some forty thousand strong, divided into four legions. This force required a similar number of non-fighting ancillaries, so some eighty thousand men, their armaments, animals and supplies had to be transported throughout the new dominion as it fell to spear and sword. This meant that the first users of the roads were the military, but soon after the conquering soldiers were established in forts and the infrastructure of empire had been set in place, the roads became essential arteries for exploiting the new lands and the indigenous peoples. Smaller roads led off the main routes and at appropriate spots new villages and towns sprang up. By the time the Roman Empire collapsed in the fifth century and new invasions followed, there was in place an intricate network of roads connecting London, as its primary hub, to a collection of important towns. These roads branched out to hundreds of villages and small towns, from Cornwall in the south-west to Hadrian's Wall in the north.

The importance of the Roman road network cannot be exaggerated for it was one of the key factors in the civilising of Europe, and historians are united in the belief that, without this network, the development of towns and cities throughout the empire would have been far slower and more haphazard. The British historian Thomas Codrington has written:

> The roads constructed during the Roman occupation do not appeal to the imagination like such remains as the Wall of Hadrian, or the ruins of an ancient city ... but when the extent and permanent nature and effect of them is considered, they may claim a foremost place among the remains of Roman work in this country. They are part of the network of roads that covered the Roman world; for many centuries they continued to be the chief means of communication within the island; and while some of them are still seen in almost perfect condition, portions of many more form part of the foundations of roads now in use.[1]

The roads even provided the means to create the first postal service, a form of Pony Express. This began as a purely military necessity. Good roads meant that horsemen could convey messages across Europe between commanders and spies on the front line and the Senate in Rome. Soon, aristocrats and politicians started to use the postal system for their personal correspondence and a booming business was created.

But, as with so many things, when the Roman Empire disintegrated the infrastructure it had created also fell apart. Fortunately, in the case of the road network this dissolution took time and the system remained pretty much intact until perhaps the seventh century. From about that time onward, though, gradually the bridges tumbled, the roads cracked and the weeds took over.

Strikingly, the roads in Britain did not again reach the standards established by the Romans until Victorian times, almost two millennia after Claudius's army had begun building its web of highways. As late as the eighteenth century the roads across Europe, and particularly in Britain, were in an appalling state and became a subject of great complaint among rich and poor. The lower classes were affected by the terrible state of the roads because it hindered trade, and the rich, who travelled widest and most often, were socially encumbered by the lack of a decent road system. According to some sources, as late as the mid-eighteenth century the journey from London to Edinburgh, a distance of some four hundred miles, took two weeks, and at around the same time the emminent politician Lord Hervey was able to write to a friend with the warning that 'The road from this place [Kensington] is grown so infamously bad that we live here in the same solitude as we would do if cast on a rock in the middle of the ocean; and all the Londoners tell us that there is between them and us an impassable gulf of mud.'[2] In those days Kensington was a village separated from London by open countryside. It lies no more than a mile from Hyde Park Corner and is today considered central London.

The difference between the achievements of the Romans two thousand years ago and early Industrial Age societies is striking, and the reason for this disparity comes down to a matter of military imperative. The roads of Britain, indeed those stretching across much of the populated world, only began to approach, and eventually to surpass, the standards of the Romans when a new empire, the British imperial state, began to expand in the wake of the Industrial Revolution. Before this time, British monarchs and the leaders of European nations made do with the crumbling infrastructure they had inherited. The British fought their wars abroad and might even have considered the building of a good road system to be dangerous as it could help invaders to spread their forces throughout the islands. In mainland Europe, military forces involved in all the wars between the decline of Rome and the rise of Napoleon had to move their men, equipment and supplies along perilous routes that were overgrown and decayed through centuries of neglect.

One exception was the road between England and Scotland. This, like all the other routes originally constructed by the Romans, had been allowed to fall into disrepair, but the English needed to move their armed forces quickly and efficiently to northern regions so they could defend themselves against the warlike and vengeful Scots who resisted any form of English rule. This road became particularly important at the beginning of the eighteenth century when successive rebellions north of the border threatened to destroy the Union of Scotland and England created in 1707.

During the 1720s, the English government appointed General Wade to organise a road-building programme that would speed movement of troops marching north to protect the border. This programme worked to the great advantage of the English until the Jacobite Rising of 1745, when the rebel forces used the same road to enter England with a large force and came within 120 miles of London.

The Victorians adopted the mantle of road builders. With money and imperial ambitions they employed the new technologies that came out of the Industrial Revolution. They also followed the guidelines laid down by the Romans and built along many of the existing routes. Their motives were identical to those of the Roman emperors – the need for a transport system for their own growing empire, to move troops and all that was necessary to occupy and exploit. As this was happening, in an effort to keep up with the onward march of British imperialists, in mainland Europe

France, Germany, Russia and Italy were all spending large amounts of money and employing thousands of men in an effort to build a new infrastructure, one which soon started to crisscross the entire continent.

Meanwhile, in the United States, first the American-Mexican War of 1846–8 and then the Civil War precipitated vast and far-reaching road-building programmes. The Army Corps of Engineers was responsible for building many of the roads and bridges that allowed the Union Army to transport men and supplies to the battle fronts. One of their most famous efforts was the building of a 2170-foot pontoon bridge across the James River in June 1864, the longest floating bridge erected before the Second World War.

However, as much as the building of roads and bridges aided enormously the merchants and traders who followed the redcoats into far-flung parts of the world, or those who plied their trade after the end of the Civil War in America, this revival was coming precisely at the time when an entirely new form of transport was creating an enormous impact. As workmen laid stone upon stone and compacted them with gravel and sand threading their paths across the civilised world, new, metal tracks were never far behind.

The Arrival of the Railway

The idea of using rails came long before the invention of the train. As early as 1550, wooden tracks were being used to carry primitive trams over waterlogged land, and by the middle of the eighteenth century, in the tin mines of Cornwall and at the coalfaces of northern England, pit ponies were being used to pull carriages along rudimentary metal tracks.

The train itself was one of the most important offspring of the Industrial Revolution and the man credited with its invention is George Stephenson, an engineer who was born near Newcastle upon Tyne, in the industrial heartland, in 1781. Stephenson worked at a coal mine where rudimentary rail tracks were used and he became interested in the new technology of steam power. In 1814 he built his first steam locomotive; it was used to haul thirty-ton loads of coal at a top speed of 4 mph. Within fifteen years he had designed and built the first passenger train, pulled by the *Rocket*, which ran on the embryonic rail network that was already in use as an industrial transportation system.

Realising the commercial potential of passenger trains, the British government enthusiastically subsidised the building of track and the development of improved steam engines. By the mid-1850s, just a quarter of a century after the maiden journey of

Stephenson's *Rocket*, more than eight thousand miles of track had been cut through the British countryside laid by an estimated 250,000 workmen who acquired the nickname 'navvies', derived from the word 'navigator'.

An indication of the changes produced by this gargantuan construction programme may be seen in the way it affected the cost of travel. In 1800, the price of a return ticket between London and Manchester was £3 10s. Fifty years later, at the time of the Great Exhibition of 1851, the same journey by train was much faster and cost just 5s. (one-fourteenth the cost).

However, not everyone was enamoured with the train. In the 21 March 1825 edition of the *Quarterly Review* an article appeared in which the author declared:

What can be more palpably absurd and ridiculous than the prospect held out of locomotives travelling twice as fast as stage coaches! We should as soon expect the people of Woolwich to suffer themselves to be fired off upon one of Congreve's *ricochte* rockets as trust themselves to the mercy of a machine going at such a rate. We will back old Father Thames against the Woolwich Railway for any sum. We trust that Parliament will, in all railways it may sanction, limit the speed to eight or nine miles an hour, which we believe is as great as can be ventured on with safety.[1]

In his biography of George Stephenson published in 1859, Samuel Smiles detailed the protests that met the earliest plans to construct railways in Britain:

Pamphlets were written and newspapers were hired to revile the railway. It was declared that its formation would prevent cows grazing and hens laying. The poisoned air from the locomotives would kill birds as they flew over them, and render the preservation of pheasants and foxes no longer possible. Householders adjoining the projected line were told that their houses would be burnt up by the fire thrown from the engine-chimneys, while the air around would be polluted by clouds of smoke. There would no longer be any use for horses; and if railways extended, the species would become extinguished, and oats and hay unsaleable commodities. Travelling by road would be rendered highly dangerous, and country inns would be ruined. Boilers would burst and blow passengers to atoms. But there was always this consolation to wind up with – that the weight of the locomotive would completely prevent its moving, and that railways, even if made, could never be worked by steam-power![2]

Although Britain had a lead on other nations thanks to the fact that the Industrial Revolution had started there, other countries were not slow to see the benefits of a railway system. In America, the first lines appeared during the 1830s. Within twenty years, nine thousand miles of track had been laid and by 1870, five years after the end of the Civil War, this figure had grown to thirty thousand. All of this came as a direct result of the important military role played by the railways.

In all industrialised nations the construction of railways, ostensibly for the benefit of the public, was in fact driven by a military imperative. After some initial resistance from a few of the older generals and fading politicians of the day, governments quickly grasped the concept that a carefully maintained rail network could be used to move troops and equipment far more efficiently than the traditional methods. Most enthusiastic were the Germans who began an ambitious programme to lay thousands of miles of line.

Beginning around 1852, the German government sponsored the building of eleven new railway lines between major cities and the border with France to the west. A few years later, they did the same thing in the east, expanding to the Russian border. This development soon aroused the suspicions of the French and the Russians who were not slow to note that the rail networks near their borders seemed far more extensive than was necessary to meet the needs of these sparsely populated regions. French fears were confirmed in 1853 when a leaked German document entitled *A Study for an*

Expedition against Paris by way of Lorraine and Champagne reached the French government. Clearly an indication of aggressive intent, it prompted one particularly astute minister to declare in the French parliament that such a document '. . . could hardly be regarded as indicative of a sentiment of fraternity'.[3]

For the Germans this was merely the first step in a grand plan and they did not stop with the rail networks skirting their borders to east and west. As these lines were being constructed, the government was putting the final touches to a scheme to build a system of military railways in German colonial territories as far afield as Africa and the Far East (most especially Kiantschou), all with the intention of providing an infrastructure for a planned challenge to the global reach of the British Empire. Between 1860 and 1900, the Germans laid thousands of miles of track on three continents just as they were pouring money into building up their army and navy. The purpose of this build-up was to create an interconnecting system, that would eventually link the railways inside Germany to lines spreading east across India and into the Far East and south to Cape Town, a plan that was only thwarted by the British who held key regions, and, alerted to the danger, were able to block the proposed links before they could be made.

This German jingoism during the second half of the nineteenth century had enormous repercussions and it played a crucial role in determining the way the world was to change during the twentieth century, militarily, politically and socially.

The first test for Germany's new infrastructure was the Franco-Prussian War of 1870–1 in which the professional, highly disciplined and well-prepared German army, operating with clockwork efficiency, crushed the French. This war had far-reaching repercussions. First, it led to the deposing of Napoleon III, who had led his country to defeat, a move that altered radically the political structure of France. Second, the war created a united Germany, a nation that was far more optimistic and wealthy (after being paid $1 billion, equivalent to perhaps $50 billion today, in reparations by the French), and together these changes planted the seeds of competition that resulted in the First World War.

Eighteen seventy-one may be seen as the year when German expansionist ideas began to escalate, when the Kaiser could realistically conceive of a German conquest of Europe, the year Germany began to present a genuine challenge to the global domination of the British Empire. That year, one of the great architects of the German victory, the Prussian field marshal Helmuth von Moltke the elder, declared: 'Build no more fortresses, build railways', and the enemies of his country took careful note.[4] After the military disaster of the Franco-Prussian War, the French government began to divert resources to creating a comprehensive rail network that would serve civilians as well as the military. The British did the same, and to the east of Germany the Russians started to expand their own rail network. By 1895, they had 22,000 miles of track, a phenomenal development considering that just twenty-eight years earlier, in 1857, when Britain, France and Germany already possessed quite extensive networks, there was only six hundred miles of railway track in the whole of Russia.

The immediate effect of this growth was a short but brutal conflict – the Russo-Japanese War of 1904 – which was, in large part, sparked by Japanese fears over the sudden and clearly military-led expansion of the Russian rail network. The flashpoint was the mineral-rich territory of Manchuria through which, by overcoming weak Chinese resistance, the Russians had succeeded in pushing their Trans-Siberian Railway; such a strategy offended and alarmed the Japanese who saw themselves as the guardians of Asia.

During the forty-three years between 1871 and 1914, the length of rail track in Europe almost trebled, from approximately 65,000 miles to a little over 180,000 miles. Government financial support for this scheme was forthcoming for one overriding reason: politicians and military leaders realised that the rail network could be used to turn a profit during peacetime and be made readily available to the military during times of war.

This is precisely what happened at the start of the First World War. All commercial use for the rail networks of Britain, France and Germany was terminated and by the end of 1914 the French had in service more than 10,000 trains to carry their armies to the war fronts. In Germany, 2,070,000 men, 118,000 horses and 400,000 tonnes of supplies were transported in 20,800 trains. When the British Expeditionary Force arrived on the French coast in August 1914, more than 1000 trains were used to take the men to the battle zone. From there, on each side, specially designed war trains running on makeshift track each day conveyed an estimated 12,000 men from the camps to the trenches.[5] This was seen as such a fantastic effort that an editorial in the *Journal des Transports* of 30 January 1915 could proudly boast: 'One can justly say that the first victory in this great conflict has been won by the railwaymen.'

Yet, some military historians have claimed much more. A.J.P. Taylor proposed the theory that the railways were not just a deciding factor in the outcome of the Great War but that the race between the rival nations to create bigger and better rail networks than their neighbour was one of the key factors in precipitating the conflict. 'The First World War had begun,' he wrote, '. . . imposed on the statesmen of Europe by railway timetables. It was an unexpected climax to the railway age.'[6]

The development of railways throughout the world is another example of the way military ambition, political drive and nationalistic fervour push innovative thinking, free resources and bring together often disparate elements to remodel the very structure of society. All of which means the development of the railways may be considered as important as the competition between European nations to create more powerful navies.* It also bears comparison with the more contemporary race between the United States and the Soviet Union to conquer and exploit space.

---

*See Part 6.

The Age of the Car

In any reckoning of the most influential and socially important inventions in history, the car would certainly come in the top ten. Early in the twenty-first century, the car stands at the centre of a massive global economic web estimated to employ tens of millions of people and involving profits in the hundreds of billions of dollars annually. Each year, some thirty million new cars take to the road, and, according to some estimates, there are now almost one billion motor vehicles on the planet. Such ubiquity is regarded by many as one of the worst catastrophes to afflict human society. Those against the car focus on the environmental damage it causes, the number of people killed and injured by these machines and the ugliness of tarmac smothering the natural landscape. Those in favour point to the fact that the car has liberated the vast majority of people in the industrialised world, it has expanded business exponentially by providing fast road transport and it stands as a testament to the ingenuity and innovation of the engineer, the designer and the scientist.

The notion of moving a vehicle without a horse or an ox occurred to many thinkers throughout history. Roger Bacon visualised the car and wrote about it in *Opus Majus*, and Leonardo da Vinci drew a design for a self-propelled tank that bristled with guns. He and other Renaissance thinkers considered the idea of horseless carriages, but such concepts were no more than that: considerations, ideas expressed on the page and quite impractical for their time. In his *Principia Mathematica* of 1687, Isaac Newton also described a way in which a vehicle could be propelled by the 'rearward direction of steam'; it was not until the Industrial Revolution, however, that a useable source of power was discovered to fulfil the dream of mechanical locomotion and enable the invention of steam-driven cars. An American, Oliver Evans, took out the first patent on what we would now call a steam car in 1792. He called it the *oruktor amphibolos* and it was enormous; thirty feet in length, it weighed fifteen tons.

Soon after electricity was discovered and the first batteries were devised, inventors began to apply new technology to power a vehicle. A Scotsman named Robert Anderson built the first 'electric car' in 1839, but it was not a great success, primarily because the life of the battery was so short and the power it generated was insufficient to move anything heavier than a lightweight chassis. The arrival of primitive long-life batteries towards the end of the nineteenth century encouraged the enthusiasts, and by the 1880s this power source had been developed to the point where it could be used to keep a fleet of taxi cabs on the streets of London. But these vehicles were really little more than a novelty for the technology-obsessed Victorians. They could only operate over short distances and had to be recharged frequently at conveniently located stations.

In its earliest days, the horseless carriage was largely ignored by financiers from the government and by buyers from the military because few people could see any value in them. The horse was a strong, loyal and reliable beast and military men had used them for thousands of years. But a factor of equal importance was the overwhelming success of the railways. Just as the first real automobiles were being produced towards the end of the nineteenth century, the steam train was at its height, transporting millions of civilians along hundreds of thousands of miles of track, and it has also become established as a mainstay of long-distance military transportation.

Car enthusiasts took little notice of the doubters though, and continued to conduct self-funded experiments and develop prototypes financed by wealthy patrons. Ransom E. Olds, who invented and manufactured the eponymous Oldsmobile, one of the first cars in America, described the advantages he believed the car had over the humble horse in an article in *Scientific American*: 'It never kicks or bites, never tires on long runs,' he declared, '. . . and it never sweats in hot weather. It does not require care in the stable and eats only while on the road.'[1]

The most important breakthrough in the evolution of the car was the internal combustion engine which made the vehicle a serious proposition for financiers and elevated it from a curiosity to a practical machine. Two inventors of the 1860s, Etienne Lenoir in France and Siegfried Marcus in Austria, working independently, succeeded in building a workable internal combustion engine, but neither man pushed the invention beyond a toy. Nor did Lenoir or Marcus patent their devices or try to install the engine into a full-size vehicle. Indeed, Lenoir died broke in 1900 and the idea of an internal combustion engine languished until two German inventors, Carl Friedrich Benz and Gottlich Wilhelm Daimler, again working independently, built the first commercially available cars powered by an internal combustion engine.

Daimler and Benz should rightly be considered the true fathers of commercial car manufacture, and, for many, Daimler's four-wheeled passenger carrier powered by a front-mounted internal combustion engine was the first 'real' car, the first practical, saleable motor vehicle.

In Part 2 we saw how the production line and the concept of mass production had come from a military wellspring – the need to standardise and produce large quantities of weapons cheaply and quickly. However, with the advent of the car the concept of mass production re-entered civilian life and in so doing it transformed the car from a rich man's plaything to an icon of popular utility. The man who almost single-handedly made this possible was the American inventor and engineer Henry Ford, who took the small-scale engineering innovation of Daimler and Benz in Germany and transformed it into a multinational, global industry.

Born in the Midwest of America in 1868, Ford was the son of an Irish immigrant farmer. As a young man he developed a powerful work ethic and when he was sixteen his innate ambition inspired him to walk to the nearest big city, Detroit, to find work. Within a few years he had acquired considerable skill as an engineer and had begun working for the Edison Illuminating Company. In 1893, when he was just twenty-five, Ford became Edison Illuminating's Chief Engineer, and while maintaining this high-powered career he found the time and financial resources to begin designing a car that used an internal combustion engine similar to those built by Daimler and Benz.

Ford's rise to celebrity and his acquisition of a vast fortune did not happen overnight. Before he could launch the Ford Motor Company in 1903, he spent ten years raising the capital to start his first modest car plant and in designing his earliest vehicles. However, once he entered the commercial arena and started producing cars, nothing held him back.

Henry Ford's big idea was not the car itself but the way it was manufactured. Until the Ford Motor Company set up shop, all cars were handmade by craftsmen and engineers. Indeed, at the time Ford's revolutionary company gained its trading licence there were almost two thousand different companies making cars in the United States alone and perhaps the same number in Europe. A clear indication of Ford's impact is that by 1929 (by which time Ford had become a household name) there were just forty-four car-makers still in business.

Ford's methods relied upon the system tried and tested by Eli Whitney half a century earlier and known as the 'American System of Manufacture'. Somewhat reminiscent of the method of producing guns and other armaments, Ford cars were made by a large team of workers who held stations on a conveyor belt as the car moved through the stages of manufacture. Each worker performed a simple but essential task in the manufacture of the vehicle. The system was enormously efficient compared with the old methods of small teams of skilled craftsmen producing one vehicle at a time from beginning to end. By the mid-1920s Ford was turning out ten thousand of his standard vehicle, the Model T, every twenty-four hours, and in 1927, the 15 millionth Ford rolled off the production line.

Henry Ford's methods were controversial, and, although he was a powerful self-publicist and stuck to his convictions, his business practices came under attack from many quarters. He was opposed by the unions, despised by the anti-car lobby and lambasted by those who believed (with good reason) that Ford was dehumanising workmen and turning them into automata.

Whatever the arguments about Ford's methods, there is no denying that he revolutionised car manufacture and that he was also responsible, more than anyone else, for the significant number of great social changes wrought by the car. His creation led to a massive rise in the popularity of car ownership, the creation of one of the biggest industries in the world, the reliance of the West upon Middle Eastern oil, damaging quantities of pollution and the deaths of many millions of drivers, passengers and pedestrians. At the same time, Ford provided enormously enhanced freedom and vastly expanded business opportunities to many millions around the world. Ford also did a great deal for the evolution of the car. By making it one of the most popular material possessions for anyone living in the industrialised world, he provided the impetus for ongoing technological improvements to the vehicle.

Tanks, Jeeps and the Arsenal of Democracy

The idea of using motorised vehicles in war may be traced back to the very earliest days of the horseless carriage. A great pioneer of the steam car, the French engineer Nicolas-Joseph Cugnot, built an experimental moving gun platform at the Paris Arsenal in 1769. But, because this vehicle had only a top speed of 2.5 mph and had to stop every ten minutes to build up steam, it did little to impress military planners and financiers and came to nothing.

Later, though, the same principle, the idea of mounting a gun on a moving platform, did succeed. Having evolved from a machine built in 1899 by the Englishman Frederick Simms, whose 'motor-war car' boasted a Daimler engine, armour plating and two mounted machine guns, the tank was first used successfully by the British Army at the Battle of Cambrai in November 1917.

Born the same year as Henry Ford and Henry Royce, Frederick Simms is said to have coined the word 'petrol', and he has recently been recognised as one of the most important innovators in the early days of the motorcar. Unlike Ford, who was a determined and outspoken pacifist during the First World War, Simms was quick to appreciate the potential of his machines as weapons. He knew of Cugnot's efforts and was aware that a few primitive steam-driven devices running on caterpillar tracks had been used some half a century earlier during the Crimean War. He also understood the shortcomings of these devices. He believed a wheeled vehicle mounted with guns could work but that a steam engine powerful enough to move such a vehicle across rough terrain at a reasonable speed would be too large to be practical. He also understood that a practical tank needed to have a powerful engine and armour plating to protect the crew.

Simms had many of the right ingredients in his designs but he was simply too far ahead of his time. When he first proposed his devices and demonstrated them to the military between 1902 and 1905, his ideas received short shrift. Britain's War Minister of the time, Lord Kitchener, remarked that Simms's machine was nothing more than '. . . a pretty mechanical toy'.

It took the impetus of pan-European war in 1914 to motivate the more forward-thinking factions within the armed forces. By then, an American, Alvin O. Lombard, had devised a practical vehicle that moved on crawler tracks and was powered by an internal combustion engine. Two senior members of the British military establishment, Ernest Swinton and Maurice Hankey, heard of the vehicle and believed it could offer a great deal to the army. After strenuous efforts, they arranged a demonstration for Winston Churchill, then First Lord of the Admiralty. Suitably impressed, Churchill succeeded in persuading the Prime Minister, David Lloyd George, to obtain funds to develop the first tank. This resulted in the establishment of 'The Landships Committee' which commissioned a Lincoln-based company, William Foster Ltd., to produce a small armoured and armed vehicle which was initially dubbed a 'landship'. Built in great secrecy, before the machine reached the trenches of France it had already been nicknamed 'the tank' by Ernest Swinton who at various times had considered the weapon to be a 'shapeless mass' or 'a giant slug'.

Between the wars, the car grew increasingly popular and the design of commercially available vehicles improved enormously. This was in no small part thanks to lessons learned at the front. The First World War provided many new technologies and innovative methods for the car industry, including the electric starter, the use of sophisticated engine lubricants, the introduction of antifreeze and techniques to improve the refining of oil to make purer, more efficient fuel. This new level of importance placed on mechanised units and road transportation led to an expression current at the time that the Allies had 'floated to victory on a sea of oil'.

In terms of human resources, the First World War made an enormous impact on the skills of mechanics, designers and engineers, many of whom had found ways to improve mechanised units and to outwit the enemy while labouring under appalling conditions and tremendous stress. The result was a wealth of experience for those who survived the war on both sides, experience that was then passed on to the commercial sector after the end of hostilities.

In the Second World War, the car and a range of armoured vehicles were indispensable to both sides. Entire mechanised units, created between the wars, became instrumental in almost every campaign between 1939 and 1945. Eighty times more fuel was used during the Second World War than in the First. At the height of the hostilities, fifty million tonnes of petroleum per week were shipped from America to Europe, and it was estimated that to keep a single mechanised division on the move required more than 18,000 gallons of petrol per hour.

Following the Second World War, the car went through another series of improvements even more profound than the changes following the Great War, thirty years earlier. The fluid drive transmission which became universal from the late 1940s was developed during the early days of the war. Synthetic lubricants and new forms of antifreeze called polyglycols, first employed in aircraft engines to make them run more efficiently, were later adapted for mechanised units all over the world before becoming an essential for all cars after the war. By the 1950s, DuPont, Sun and Mobil had each introduced synthetic hydrocarbon oils that not only made cars run far better but greatly improved the longevity of the engine.

Between 1942 and 1945, the American government spent $20

billion on updating manufacturing and converting car factories into munitions plants to mass-produce tanks and aircraft. When the war was won, these plants were taken over by the car makers and they became integral to the massive expansion of the motor industry after 1945. The boom in the industry brought cheap, easily obtainable cars to the market so that the average person living in the industrialised world could travel more freely than ever before and at a reasonable cost. Henry Ford's Model T, manufactured during the 1920s, had precipitated a shift in the market for cars. With America making vast fortunes from the Second World War, the economy was boosted enormously, so that for the first time working people had far greater disposable income and the sale of cars accelerated rapidly.

As a reaction to the austerity of the war years and because of new American wealth, the late 1940s saw a proliferation of innovative vehicle designs. This progress was particularly noticeable in the evolution of the truck from a heavy, slow and often rather unattractive vehicle before the war to much sleeker, faster, more efficient models. From about 1947 onwards, these new designs were embraced by businesses across America and Europe and led to the introduction of huge trucks and the deployment of fleets of articulated lorries.

Parallel with the development of the truck, the jeep made its first appearance in the civilian world immediately after the war. An American firm called Willys cornered the market, successfully crossing over from military supplier to commercial retailer, and became the leading producer of hardy, versatile vehicles for the Midwestern farmer, with tens of thousands of machines rolling off the assembly line. Later, other large car manufacturers began to dominate the market, including Ford and Chrysler. Today's four-wheel-drive SUVs (sports utility vehicles) are the descendants of the jeeps and armoured cars of the Second World War, vehicles that first employed all-wheel drive and off-road mechanical systems to operate in all weathers and across all types of terrain.

Although a pacifist during the First World War (even financing a so-called Peace Ship that was intended to sail to Europe to help broker an armistice), by the time America had entered the Second World War, late in 1941, Henry Ford had become a supporter of the fight against Germany and Japan. His expertise and experience were invaluable and his high-profile involvement had the effect of pushing other car manufacturers to compete.

Turning to the manufacture of armoured cars, tanks and planes, Ford, Chrysler, Chevrolet and other companies transformed America into what was dubbed at the time 'the arsenal of democracy'. By 1942, Henry Ford could boast that he had turned the building of a B-24 bomber from a thirty-six-man, 1500-hour job to a three-man, twenty-six-hour one. During the three and a half years America was in the war, Oldsmobile produced forty-eight million rounds of ammunition and 350,000 parts for aircraft engines, and Buick manufactured 1000 aircraft engines per month. Willys built and shipped almost 400,000 jeeps, Chrysler produced 20,000 tanks, and Ford, the most prolific of them all, supplied the armed forces with 93,000 trucks, 8600 B-24s, 3000 tanks, 13,000 amphibious vehicles and almost 300,000 jeeps.

But the practical effects of the war on design and technology are only one aspect of the changes conflict brought to the automobile industry. The rise and rise of the oil and petrol companies began with the vast fortunes they made during the Second World War and precipitated their transformation into the monolithic entities they are today. In the 1940s, the American corporation Exxon and the Anglo-Dutch giant Shell began to dominate the Western market, eventually making billions of dollars annually. Yet, while the car and the fuel needed to keep the car on the road have provided millions of jobs and have become pillars of the global economy, it has also made the West vulnerable. Western society relies on oil; it is the lifeblood of civilisation. It has become a new and crucial factor in the complex mesh of global politics, an irreplaceable commodity which has forced the West to do everything in its power to maintain its flow.

This role played by the car, as an integral component in Western civilisation, began immediately after the Second World War, and it came at precisely the moment the old world order was being replaced by the new. The car has become the central pillar that supports American hegemony, which today remains unbowed, if bloodied. An indicator of this situation is the need for image building in the face of Cold War anxieties reflected by American car design. From the late 1940s through to the mid-1960s, cars became bigger, bolder, brighter and more ostentatious in a bid to demonstrate American superiority, wealth and technical prowess. Rather than fighting with nuclear weapons, East and West challenged each other using space exploration, notably in the race to the moon; but the Cold War was also fought within the minds of the public using other, more subtle instruments. The elongated tail fin, the increased horsepower and the bright red paint of 1950s cars offered some of the most eye-catching symbols of this conflict. Almost immediately after the propaganda war was won, just as the space programme was dismantled and nationalistic point scoring was replaced by concerns over the economy, oil depletion and pollution, car design shifted back to a more prosaic level.

Mankind started by waging war on foot. Later, the horse and the ox were employed and then mechanised vehicles were developed for commercial and military purposes. Some six millennia after adopting the horse, society has been fundamentally altered by the technology we use to get around, to run our businesses and to move our armies. We have come full circle: we no longer drive the car, the car drives us.

PART 5

FROM

THE BALLOON

TO THE

SPACE SHUTTLE

Mankind has always yearned to fly. The ancient Chinese mastered the art of kite making probably around 1000 BC, using bamboo and silk. Kites were originally used for festivals and pageants, and one famous craftsman named Kungshu Phan built a bird-shaped model that was reportedly kept aloft for three days. But by about 400 BC the ingenious Chinese had adopted kites for military purposes, employing them as a distraction immediately before attacking an enemy.

In the West, legends tell of the fascination with flight. The myth of Icarus who flew too close to the sun and plummeted to earth after the wax attaching his wings melted is more a metaphor than anything based upon reality. However, there were almost certainly many attempts to conquer the air in ancient times. The tale of Daedalus, for example, who escaped from the tower of King Minos by making a pair of wings, may have been based upon ancient stories of brave (or perhaps insane) souls who tried to find methods of flying like birds.

Ancient texts give vague descriptions of possible flying machines. A fourth-century AD Chinese text called *Pao Phu Tau* mentions just such a device: 'Someone asked the master about the principles (tao) of mounting to dangerous heights and travelling into the vast inane. The Master said, "Some have made flying cars with wood from the inner part of the jujube tree, using ox-leather straps fastened to returning blades so as to set the machine in motion."'

Medieval and Renaissance innovators took the idea of flying very seriously. Roger Bacon was stifled by convention and probably never went as far as actually building any form of flying machine, but he wrote extensively on the possibilities of flight and studied the notion of buoyancy, proposing that there was 'solid' material in the air that could keep heavier-than-air objects aloft if they were designed in the proper way. In his *Epistola de secretis operibus naturae, et de nullitate magiae* (*Book of Secret Operations and Natural Magic*) Bacon wrote:

> First, by the figurations of art there be made instruments of navigation without men to row them, as great ships to brooke the sea, only with one man to steer them, and they shall sail far more swiftly than if they were full of men; also chariots that shall move with unspeakable force without any living creature to stir them. Likewise an instrument may be made to fly withall if one sits in the midst of the instrument, and do turn an engine, by which the wings, being artificially composed, may beat the air after the manner of a flying bird.

Leonardo covered hundreds of pages of notebooks with ideas about flight and the pages that survive show that he had some advanced ideas about how to design a flying machine. 'There shall be wings!' he declared in the *Codex Atlanticus* of about 1480, now in the Ambrosiana Library, Milan. 'If the accomplishment be not for me, 'tis for some other. The spirit cannot die; and man, who shall know all and shall have wings.' And although Bacon had also described a similar machine around 1250, Leonardo's ornithopter (or helical air screw) may be thought of as the ancient precursor of the helicopter. 'I have discovered that a screw-shaped device such as this,' he continued, '. . . if it is well made from starched linen, will rise in the air if turned quickly.'

Working models of Leonardo's designs were almost certainly never constructed and, like his descriptions of tanks and other ground vehicles, they remain nothing more than exquisite drawings and carefully thought-out plans. As with so many of his efforts, Leonardo was stymied in his attempts to build a practical flying machine not so much by design conundrums as by the limits of the technology available to him. Most importantly, there was no source of power available other than human muscles, horse or oxen and water power, none of which were very useful to the determined aviator. Some early innovators believed man could fly using muscle power alone, but such means actually provide only a fraction of the power needed to keep a human being aloft and our bodies would be too heavy for such a purpose even if we attached huge wing extensions to our arms.

Leonardo seemed unaware of this drawback, writing in his *Treatise on the Flight of Birds* (written around 1505 but never published): 'A bird is an instrument working according to mathematical law, which instrument it is within the capacity of man to reproduce with all its movements.' The hypothesis that humans could never fly under their own power was proven in 1680 by Giovanni Borelli, a medical man and flight enthusiast. He showed that human muscles were too weak to move the large surface area needed for effective wings and that in order to provide the necessary power the heart would need to beat at a rate of eight hundred beats per minute.

It is fair to say that before the eighteenth century most attempts to fly were misguided in the extreme. Almost all experimenters were drawn to the notion of replicating the behaviour of birds and attempting to build machines that could imitate the flapping action of their wings; they were doomed to failure. Gliders were built with some success but they were dependent upon the capricious nature of the wind. The first successful method of flight came about by the application of lateral thinking: the development of a lighter-than-air vehicle that made no attempt to copy birds at all.

Up, Up and Away

One day in 1782 a forty-one year old Parisian papermaker named Joseph Montgolfier was sitting at home in front of his fire contemplating the humiliating military situation in Gibraltar. The British had taken the tiny island off the south coast of Spain and the French had since found the garrison to be impregnable by land or sea. As he reflected on this situation he watched one of his shirts that had been put out to dry on a stand in front of the fire. Suddenly, a puff of smoke from the fire billowed the shirt. It came free from the stand and rose to the low ceiling of his living room. It was this serendipitous occurrence that gave Montgolfier the idea of building a balloon that could be used by the French to attack Gibraltar from the air.

This at least is the story behind the inspiration for Joseph and his brother Etienne Montgolfier's involvement with balloons. Just like many of the great stories of invention and discovery, this tale may have some foundation in fact, but it could just as easily be a neat fiction created after the facts by one of the Montgolfiers or by a friend. Whatever the spark, Joseph and Etienne were soon experimenting with paper bags filled with steam before moving on to paper containers filled with hydrogen produced by mixing sulphuric acid with iron filings. Neither of these sets of experiments proved very successful. However, on a third attempt, this time using a taffeta envelope in place of Joseph's shirt, the brothers made it rise to the ceiling using smoke from a small fire.

It is a testament to the tenacity and determination of the Montgolfiers that within a year they had graduated from making a small taffeta bag float to the ceiling of their living room to launching the first manned balloon. There have been claims that others devised the hot-air balloon before the Montgolfiers. Most notable was a Peruvian experimenter Bartolomeu Laurenço de Gusmão, who had apparently made a short balloon flight eighty years earlier but had failed to patent his work. This leaves the Montgolfiers to be acknowledged as the inventors of a genuine method of sustained flight in a lighter-than-air craft.

To put this achievement into perspective we should consider the attitudes of many of the Montgolfiers' contemporaries. A little over a year before the first manned balloon flight, another Parisian, an academic named Joseph de Lalande, was quoted in the *Journal de Paris* as saying: 'It is entirely impossible for man to rise into the air and to float there. For this he would need wings of tremendous dimensions and they would have to be moved at three feet per second. Only a fool would expect such a thing to be realised.'

Although neither of the Montgolfier brothers was a scientist their business background as papermakers was useful to them. They had the money to experiment and they knew some influential people. Most importantly, their balloons were made from paper, linen and silk lined with alum for fireproofing. The sections that made up their first successful balloon, an envelope thirty-eight-feet tall, were kept together by more than two thousand buttons; these were materials they worked with every day.

The first true flight was demonstrated on 19 September 1783 at the Palace of Versailles outside Paris, specially for King Louis XVI. The balloon was unmanned but carried three passengers: a sheep, a duck and a cockerel. It travelled a distance of two kilometres and reached a maximum height of some six thousand feet. The sheep and the duck survived but the cock broke its neck on landing.

Two months later, on 21 November, the first manned flight took place. Two men, a physicist named Jean-François Pilatre de Rozier and an aristocrat, François Laurent, the Marquis d'Arlandes, boarded a new and much-improved Montgolfier balloon. During their twenty-minute flight they travelled eight kilometres, and their courage made themselves and the creators of the first flying machine national heroes. Indeed, Joseph and Etienne became so famous that for some time thereafter a balloon was known simply as a 'Montgolfier'.

Within six months the brothers were honoured by the French Academy of Sciences as '. . . scientists to whom we are indebted for a new art that will make an epoch in the history of science'. This prediction quickly proved to be accurate. Almost immediately, ballooning became a sensation as a sport, a pastime for the adventurous and wealthy, a tool of science, and, crucially, a most valuable instrument in the hands of the military.

It is a remarkable fact that the Montgolfiers had no real idea how their balloons lifted off the ground. They were under the false impression that it was the smoke from their furnaces that propelled their devices. Rather, it was the principle of buoyancy. The air inside the balloon is heated and expands, so that in a given volume there are fewer molecules than in cool air. Being less dense this heated air floats above the denser cool air, pushing the balloon up with it. When the air in the envelope cools down, the balloon descends unless it is heated again from a thermal source carried above the basket and the heads of the passengers.

This explanation was formulated by the famous scientist Jacques Alexandre Charles (the creator of Charles' law of gases, learned by rote by most schoolchildren) who was also one of the first men in the world to fly in a balloon. Charles made ballooning a science and he experimented with different gases and designs, taking the Montgolfier's innovation and turning it into a practical form of aerial transport.

Although interest in ballooning spread around the world very quickly, it was the French who not only pioneered its use but led the way in all early aviation developments. French engineers and scientists were also the first to understand the military value of the balloon. Less than ten years after the first flight (a period scourged by the horrors of the French Revolution), the French army was using balloons for reconnaissance; and the first balloon built specifically for military purposes was *L'Entrepremant*, designed by the French scientist Charles Coutelle under the instructions of the Committee of Public Safety early in 1794.

L'Entrepremant and three balloons – the *Celeste*, the *Hercule* and the *Intrepide* that quickly followed it – were hydrogen balloons that were tethered to the ground by sturdy ropes and floated high above the approaches to a battle site. Communication with the ground came in the form of messages sent along the tether or by using semaphore, and during the first experimental use of *L'Entrepremant* the pilot, Coutelle himself, claimed he could see troop movements eighteen miles away. The Committee of Public Safety was so impressed with this first demonstration of a balloon as a surveillance device that in March 1794 it established the world's first air force, the Compagnie d'Aéronautiers, or Balloon Corps.

Hardly had the corps been created than it was in action. The Battle of Fleurus in June 1794, during which the French were victorious against the invading Austrians, was the first time the balloon was used in a real conflict (rather than a demonstration). Again, Coutelle was the pilot and he gave detailed reports of enemy movements at least an hour before they were visible over the horizon from ground level, a success that later led some observers to conclude that the reconnaissance balloon had been a key factor in the French victory. Indeed, the Austrians were so shocked that there were calls for the use of balloons to be banned because it broke the rules of engagement. News that their movements had been observed so far in advance led many of the uneducated Austrian foot soldiers to conclude that the balloon was the work of the devil, who had sided with the French.

Soon, the technology of ballooning spread abroad and the French lost their initial strategic advantage. Napoleon was persuaded by his advisers to take the Balloon Corps on his Egypt campaign of 1798 but it was never used because the British raided the French camp and destroyed all the balloons and associated equipment before the decisive Battle of Aboukir. Upon returning to France, a disillusioned Napoleon disbanded the Balloon Corps. This had the long-term effect of slowing the advance of aviation by at least two generations, and, in the short-term, of destroying the dreams of a small band of enthusiasts within the military establishment who had been working for years on a proposal for a balloon-led invasion of Britain.

Fifty years later there was a revival in the military use of balloons. By this time balloon technology had improved considerably in the hands of enthusiasts who had adopted it for sport and leisure. Scientists, too, had made great use of the balloon for observation, weather testing and for studying the atmosphere. By the 1860s, the balloon was ripe for a renaissance, thanks to a conflict far from Europe, the American Civil War.

Two men were responsible for this resurgence of interest. The first, Thaddeus Lowe, was an entrepreneur, showman and egotist who seemed to clash with everyone he worked with. His friend and agent, a wealthy businessman named Murat Halstead, succeeded in contacting high-ranking officials in the Union Army and convincing them of the usefulness of balloons. To enlist the support of President Abraham Lincoln, Lowe and Halstead arranged for a demonstration in the skies over Washington, DC. In August 1861, Lowe's balloon, the *Enterprise*, was tethered some five hundred feet above the centre of the city and Lowe showed that he could telegraph observations to the ground. Lincoln was suitably impressed and immediately approved funding for a small corps of surveillance balloons to spy on Confederate positions.

Lowe had secured support just in time. Another adventurer and entrepreneur, John LaMountain, had also developed a practical reconnaissance balloon but he had approached government financiers too late and had failed to secure sponsorship. Undeterred, LaMountain decided to bypass the politicians. He went directly to the commander of the Union forces south of Washington, Major General Benjamin Butler, who was then at Fort Monroe in Virginia.

Inspired by LaMountain's enthusiasm and determination, Butler allowed him to use his balloon, the *Atlantic*, in reconnaissance missions over enemy lines.

LaMountain and Lowe were both eccentric individuals and neither of them lasted long as adjuncts to the military. Lowe was an uncompromising, abrasive character who bridled at the arrival of LaMountain on the scene and refused to cooperate with his fellow balloonist. LaMountain was probably the more innovative of the pair. While Lowe kept his balloons tethered high above Union positions, LaMountain used the balloon as a vehicle as well as a stationary surveillance platform and he actually flew over Confederate lines, was shot down and escaped to report back his observations. He also pioneered further military applications for his creations including launching a balloon from a ship, a small vessel called the *Fanny*, anchored in the James River.

Neither Lowe nor LaMountain could properly do business with the military and both men lost their army contracts before the war was over. Lowe had some political influence and lasted longer, but LaMountain's legacy survived the more intact, in part thanks to good friends in the newspaper world who popularised his ballooning ideas. He died at the age of forty, within a decade of his innovative contributions to the Civil War.

Renewed interest in the balloon was not confined to the United States. In 1870, five years after the end of the American Civil War, France was fighting Prussia, and by September of that year Prussian forces had laid siege to Paris. Little could be done to break the siege or to bring much-needed supplies into the French capital, but a group of ballooning enthusiasts, inspired by the uses to which the Americans had put balloons, managed to persuade the head of the French postal service to employ them to transport mail to French neighbours and allies. By the end of the year, when the siege was finally broken, more than a hundred people had escaped the city in balloons, but, equally important, an estimated one million pieces of mail had been delivered abroad by this method.

Military analysts across Europe were inspired by the innovative use of balloons in Paris during 1870, and several countries began to set up air corps along the lines of Napoleon's short-lived Balloon Corps seventy years earlier. The French revived their air force, and in 1874 they established a Commission des Communications Aeriennes. By the end of the decade, the Russians, Germans and Austrians had their own air forces, and thanks to the perseverance of two great enthusiasts who were also army captains, Frederick Baumont (who had served with Lowe during 1862) and George Glover, the British Army was finally persuaded to establish a balloon corps of their own at the Woolwich Arsenal.

At about the same time, news of the military uses of balloons served to increase interest in their scientific application. The earliest scientific flights were made simply to study the nature of the air at higher altitudes, but by the first decade of the twentieth century there were at least a dozen different scientific organisations conducting regular experiments at high altitudes. In 1912, a scientist named Viktor Hess took an electroscope aboard a balloon to test radiation levels at different altitudes. This led to the important discovery that radioactivity increases at higher altitudes, showing that the earth itself is not the only source of radiation, and that a stream of particles is emitted constantly by the sun and brought to earth within the solar wind.

Today, balloons are used extensively in atmospheric research because they are far cheaper than planes or space vehicles. They can carry very large payloads and stay aloft unmanned for long periods. Using balloons, scientists have continued to study atmospheric conditions and weather patterns (as the first researchers did), but they can also analyse meteorite and micro-meteorite activity and they can be used to observe magnetic phenomena and other exotic atmospheric effects, all of which makes them invaluable to meteorologists.

The balloon is immensely useful to the astronomer as well. Large telescopes can be mounted on balloons, offering far clearer images than ground-based observatories. Indeed, until the launch of the Hubble telescope in 1986, balloon-mounted telescopes provided scientists with the best method of making astronomical observations unhindered by pollution and other atmospheric disturbances.

Soon after balloons were first used for wartime surveillance, they began to be employed by civilian aerial photographers. The pioneer in this field was the French balloonist Gaspard-Felix Tournachon, who in 1859 took the first high-quality aerial photographs from a balloon. As altitude records were surpassed, the aerial photographs became ever more spectacular, and for a public which had never seen such things before some of these pictures seemed truly miraculous. In 1935, Captain Albert Stevens, the pilot of a helium-filled balloon called *Explorer I*, ascended to a height of 72,000 feet (twice the height of Everest) and took photographs of the whole of South Dakota stretched out in all its glory.

A significant development of the balloon, but one that has suffered more than its fair share of bad press, is the airship. A direct descendant of the earliest balloons, the airship, or dirigible, is simply a propeller-driven balloon with an internal framework to support it; it is powered and, to a limited degree, steerable.

The earliest efforts to build an airship again came from French pioneers but these simply involved attempts to steer a gas bag. In 1884, Charles Renard and Arthur Krebs, both officers in the French Army Corps of Engineers, designed, built and flew *La France*, a powered balloon that travelled eight kilometres using an electric battery which generated seven horsepower. But the first true airship, a vessel with an inner framework that could be powered and steered, was the LZ-1, which embarked on its maiden flight in 1890 under the guidance of its designer, Count Ferdinand von Zeppelin.

Zeppelin had served in the American Civil War as a foreign attaché and he had witnessed first-hand the use of balloons in warfare. As an army officer his earliest ideas about his aircraft were concerned with how they might be applied to military usage, although he later embraced commercial and civilian uses for the machine. His first efforts were self-financed, but within a short time he had exhausted his funds and his experiments came to an end. For two years he concentrated on obtaining adequate funding, and, in 1905, he succeeded in securing a contract from the German military to build an improved version of the LZ-1, which he called the LZ-2.

By the beginning of the First World War the German Army had an entire fleet of dirigibles, or 'Zeppelins' as they came to be known. Between 1914 and 1918, the fleet conducted more than five hundred raids on France and even reached as far as London. However, they were largely ineffectual, and, compared with the horror of the Blitz a generation later, the Zeppelin raids generated more anger than any serious damage. But they caused considerable anxiety within the British military establishment, which came to believe that such technology could be developed and honed into a truly effective weapon. This belief generated competition between the military establishments of Britain and Germany, so that, by the end of the war, the British had some two hundred airships in service, making it the largest air force in the world at the time.

The First World War offered considerable impetus to airship designers and builders and between the wars Britain took the lead. The high point of the British effort came with the R-100 and R-101 which were capable of carrying scores of passengers in some comfort between London and many of the major cities of Europe. Sadly, though, this adventurous programme was cut short by tragedy. In 1930, the R-101 crashed, killing all forty-seven passengers and crew. With this crash the British commercial airship programme also died.

Germany, however, persevered with building airships and Hitler himself was said to be an enthusiast of the Zeppelin. But this programme, too, ended in tragedy. After a transatlantic flight on 6 May 1937, the *Hindenberg*, the largest airship ever built, erupted into a deadly fireball as it docked at Lakehurst, New Jersey. The crash was captured on film and this record is now considered an iconic piece of footage. By the morning after the disaster, the world's newspapers were carrying the story and public interest in airships vanished instantly.

The demise of the airship was both premature and regrettable because this means of transport still has a great deal to offer. The terrible fault with the early airships such as the R-101 and the *Hindenberg* was that they were filled with hydrogen gas, one of the most flammable substances known. Modern airships, which are beginning to find a renewed, if limited, use, are helium-filled. Helium is inert and therefore safe. Airships remain slow vehicles compared with aeroplanes, and today the aim for all means of travel is a balance between speed, safety, comfort and economy. Coming from a slower, perhaps more genteel era, the airship, trapped in its time, might have offered a far cleaner, cheaper and certainly more picturesque method of air travel. But as the early airships were failing through misfortune and poor design concepts, a new and far more adaptable means of air transport that offered the key to the future of aviation was arriving on the scene.

To Fly Like a Bird

'I confess that in 1901,' Wilbur Wright told an audience at the Aero Club of France in 1908, 'I said to my brother Orville that man would not fly for fifty years . . . Ever since, I have distrusted myself and avoid predictions.'

Orville and Wilbur were the second and third of four children born to Bishop Milton Wright and Susan Catherine Wright, Mid-Westerners who travelled around the country while their children were growing up and who finally settled in Dayton, Ohio, in 1884.

Born in 1867, Wilbur was older than Orville by four years, but the brothers were particularly close and as they entered adulthood they shared an interest in engineering and science, although neither of them attended college. In 1889, they created together a four-page newspaper called *West Side News* before setting up a bicycle manufacturing and retail business at 1127, West Third Street, Dayton. It was on these premises and using just the tools in their workshop that, beginning in 1896, the brothers designed and built a succession of craft which eventually evolved into the first heavier-than-air flying machine.

The Wrights were simple men. They understood mechanics instinctively and taught themselves both theory and practical engineering. Without a single day of formal further education between them, but by demonstrating exceptional flair and imagination, they were able to conduct precise experiments that were professionally documented and analysed. They first experimented with simple gliders to learn basic principles. Later they discovered how best to power their flying machines with engines they built themselves. They worked together but in seclusion; so paranoid were they about secrecy that this would later prove a hindrance in their efforts to acquire the commercial rewards they deserved.

The first successful heavier-than-air flight took place on 17 December 1903, at Kitty Hawk, North Carolina, on a windswept and deserted beach the brothers had been using to test a succession of craft since 1901. Orville Wright recorded the event in his journal, and, later, he and Wilbur described the scene in a magazine article:

> The first flights with the power-machine were made on the 17th of December, 1903. Only five persons besides ourselves were present. These were Messrs. John T. Daniels, W. S. Dough, and A. D. Etheridge of the Kill Devil Life Saving Station; Mr. W. C. Brinkley of Manteo, and Mr. John Ward of Naghead. Although a general invitation had been extended to the people living within five or six miles, not many were willing to face the rigors of a cold December wind in order to see, as they no doubt thought, another flying-machine *not* fly.[1]

That first flight lasted just twelve seconds and took the pilot, Wilbur Wright, ten feet into the air and a distance of 120 feet, but it was history in the making, the first time a human being had flown in a machine heavier than air and powered artificially. By the end of the day, the Wrights had made three more flights. Orville took the aeroplane a distance of two hundred feet, and, as the light faded during the last launch of the day, Wilbur travelled 859 feet and stayed aloft for fifty-nine seconds.

The Wrights conducted their research (and spent a great deal of their own money) simply because they wanted to fly. 'We had taken up aeronautics merely as a sport. We reluctantly entered upon the scientific side of it,' the brothers recalled in 1908. 'But we soon found the work so fascinating that we were drawn into it deeper and deeper.'[2] But in 1903 they were not the only ones interested in achieving this goal. Although they were not aware of his work and had never met him, the brothers had a rival in the shape of one of the world's leading scientists, Samuel Langley, who was armed not only with formidable scientific knowledge and prestige but was also being funded by the United States military.

Langley was a world-renowned astronomer working at the Smithsonian Institution in Washington, DC. In 1898, during the Spanish-American War, he had secured a $50,000 grant from Congress to design and build a steam-powered craft he and his assistant and chief pilot, Charles Manly, called an Aerodrome. It is some indication of Langley's influence that, seven years before the Wrights changed the history of aviation, Langley could persuade political figures to finance such an enormous grant (several million dollars in today's terms).

To appreciate how difficult it must have been to secure funding for projects to build flying machines we need only consider some of the things respected scientists and establishment figures were then saying about the possibility of manned flight. In 1895, three years before Langley had acquired his grant, the most famous scientist in the world, Lord Kelvin, the President of the Royal Society in London, had declared: 'Heavier-than-air machines are impossible.' The same year, in America, the great Thomas Edison had written: 'It is apparent to me that the possibilities of the aeroplane, which two or three years ago were thought to hold the solution to the flying machine problem, have been exhausted, and that we must turn elsewhere.'[3] A few years later, and only months before the Wrights flew, Simon Newcomb, professor of mathematics and astronomy at Johns Hopkins University, declared in a lecture: 'Flight by machines heavier than air is unpractical and insignificant, if not utterly impossible.' Later, he compounded his misconceptions by declaring: 'Aerial flight is one of that class of problems with which man will never be able to cope.'*

Clearly, though, there were some enlightened people in powerful positions who were also friends of Langley's and he did have considerable cachet as a famous astronomer. However, for all the money poured into his project (and the initial $50,000 supplied by the US government was just the beginning), his designs were flawed, his engines too heavy and underpowered and he was beaten to the prize of being the first to conquer the air by two young, unsophisticated men toiling away in the workshop of a family-run bicycle factory.

Oddly, even after Wilbur and Orville Wright had succeeded in their task and could offer photographs and moving pictures of their exploits, many people could not accept what they had accomplished. The Wrights were themselves partly to blame for this. They were oversecretive and they trusted very few people enough to offer them access to their carefully documented experimental findings. They were aware of the huge commercial potential of their craft and were, of course, not slow to appreciate that the military might be interested in funding their further researches. According to one historian: 'They [the Wright brothers] were able to get some early government contracts to sell planes to the military and train some of its first pilots. They felt it was important that the military get on board early.'[4] But these efforts were initially only half-hearted because for several years the Wrights refused to give demonstrations to government representatives, allowing others with fewer anxieties and more commercial acumen to build their own machines and to cash in on demand.

The reluctance of the US government to throw money at the Wrights without a proper demonstration is understandable. After losing so much money through Langley's failed device (his final

---

*This last is, in itself, an odd thing to say. After all, what is flight if it is not aerial? Recently, a group of computer professionals have established the Simon Newcomb Prize which they award to the most ridiculous published argument against Artificial Intelligence.

and most expensive Aerodrome nose-dived into the Potomac River nine days before the Wrights' triumphant flight), it is surprising that anyone in authority could be persuaded even to consider financing a flying machine.

Disillusioned with American politicians and military men, the Wrights approached the British government and offered it the patents to their aircraft. They received this reply: 'Their Lordships are of the opinion that they would not be of any practical use to the Naval Service.' Fortunately, soon after this rejection a few senior figures in Congress began to realise the military potential of flying machines, and by 1907 the now cash-strapped Wrights finally agreed to give a demonstration of their aeroplane. This led to an offer of $30,000 to supply a craft capable of carrying two men and sufficient fuel for a flight of 125 miles at a speed of 40 mph.

In the space of just three and a half years the art of flying had captured the imagination of many experimenters around the world and progress had come with lightning speed. The Wrights' first flyers were flimsy, unstable craft that could travel just a few hundred yards, but by 1908 they were constructing machines that could meet the demands of their contract with the military and they began to supply the US government with planes that could travel the distances demanded and at higher speeds than the minimum specified. By the start of the First World War, only eleven years after the Wrights' first flight, the British, Germans, French, Italians, Russians and Americans all possessed warplanes, many of which were soon to see action in France.

The British had been prompted to move with the times thanks to German and French enthusiasm for the aeroplane. They had also been influenced by a growing interest in aviation around the world. Six years before the outbreak of war, the Wrights had created one of only two companies then supplying the US military with planes. Competition had come in the shape of a firm established by an aviation pioneer named Glenn Martin who, in 1908, set up his own manufacturing plant. By the end of that year, his and the Wrights' first planes were in operation with the US Army Signal Corps. These were the first military aeroplanes in service anywhere in the world and they were expected to be used purely for reconnaissance, mirroring the role played by the hot-air balloons during the Civil War some fifty years earlier. But it was not only within military circles that the aeroplane was generating interest. The first scheduled civilian air service started in Florida on 1 January 1914, using a seaplane. The company, Florida Airlines, only lasted one winter tourist season but in that time it carried more than 1200 passengers, one at a time, on eighteen-mile, twenty-three-minute flights across Tampa Bay. At the same time, daring aviators were pushing aircraft to their limits, flying further and faster, garnering admiration and making headlines everywhere they went.

The First World War saw aeroplanes in action for the first time and military planners were impressed by the fact that these devices could be used for much more than reconnaissance. From the earliest days of the war, British, German and French fliers were engaging in dogfights over the trenches and hurling grenades from their planes as they swooped along the front line. Yet, as brave as their pilots were, aircraft had little real impact upon the outcome of the First World War.

However, between 1914 and 1918 the technology of flying advanced with incredible speed so that in a very short time planes had become stronger and more reliable. The top speed of an aeroplane doubled during the First World War, so that by the armistice they were clocking up speeds of 130 mph. This progress was due in large part to great improvements in engine design. With the advent of more powerful and efficient engines bigger and heavier planes could be built. By 1918, both Britain and Germany had built bombers capable of crossing the English Channel and attacking civilian targets.

The years between the wars offered mixed fortunes for aviators. On the one hand flying became much more common and genuine commercial airlines were established using converted bombers. On the other, little progress was made in aviation technology because of the glut of aircraft left over from the fighting and also because progress had been so accelerated during the four years of war. But by the start of the Second World War commercial interest in aircraft had grown so dramatically that they had become a fixture of modern life.

The iconic DC-3 first flew in 1936 and it is considered to be the earliest truly commercial airliner because it enabled American Airlines to make a profit. It also showed that the younger designers and builders had finally begun to make genuine advances of their own after the burst of innovation sparked by the First World War. The DC-3 could boast a 50 per cent increase in the number of passengers it could carry compared with its progenitor, the DC-2, yet it cost only 10 per cent more to operate. It had noise-deadening insulation to add passenger comfort, hydraulic pumps to raise and lower the landing gear, a far more powerful engine and a considerably stronger (and therefore safer) fuselage made from a newly developed aluminium alloy. It was a very popular plane with the public and it quickly propelled the airlines into becoming a multimillion dollar, pan-global industry.

The aeroplane had seen only limited use during the First World War, but by the time the Second World War began two decades later it had become an indispensable tool to the military establishments of both sides. By 1945, aircraft were perceived as having played a key role in the outcome of the war and were considered an integral part of any future military planning. A very clear indication of their value comes from the fact that in 1939 in the United States there were no more than three hundred aircraft in regular service; six years later, manufacturers were turning out fifty thousand planes a year and there were an estimated three hundred thousand American aircraft in service around the world. In the space of those six years, aircraft production had become the largest manufacturing industry in history.

At the start of the conflict, monoplanes had almost completely replaced biplanes (although a few remained in service until the early 1940s) and the British Spitfire and the German Messerschmitt, the most advanced aircraft in use, rivalled each other for speed and manoeuvrability. But between 1939 and 1945, British engineers made more than a thousand technical modifications to the Spitfire, facilitating greater control, increased safety and durability and increasing its top speed by more than 100 mph.

Almost all of the engineering improvements made to aircraft during and immediately after the Second World War came from Europe. There were two reasons for this. It was primarily British and German pilots who engaged in dogfights. Also, the vast majority of bombers used between 1939 and 1942 were British and German, so there was great impetus for the military of both countries to spend money and effort on improving aircraft design. The second reason was that the postwar American aviation industry was obsessed with fuel efficiency and in keeping the cost of flying low. The American airlines, which saw an enormous increase in traffic immediately after the war, were greatly concerned with making a profit and so they were slow to adopt new technologies unless they had a direct and favourable bearing on the balance sheet.

The greatest aviation advance to come from the Second World War was the invention of the jet aircraft, created independently by two men working for rival military forces. The first to propose the theory of jet aircraft was a British avionics expert named Frank Whittle, but the first jet to fly was German and was designed by Whittle's opposite number, Hans von Ohain.

Whittle had a model of a jet engine working in the laboratory by 1937 but the first British jet plane to be built, the Gloster E.28/39 did not fly until 1941, eighteen months after the German Heinkel He178 took to the air. The British were quick to catch up, however, so that by the end of the war both countries had in service twin-engined successors to the E.28/39 and the He178. The Royal Air Force's Meteor was used to shoot down the pilotless flying bombs the Germans called V1s in 1944. Meanwhile the Luftwaffe's Heinkel He 280, which had topped 550 mph in trials, saw regular service against Allied troops advancing towards Germany after the D-Day landings.

Even so, the jet played only a minor role in the outcome of the Second World War. It is a complex machine and its development was too slow to allow for mass production before the fighting ended. However, in the postwar world the jet initiated an aviation revolution by allowing aircraft builders to develop airliners that could travel long distances at high speed and at a lower cost than old-style, propeller-driven aeroplanes.

During the 1950s, the British, Americans and Russians developed the bomber to the point where it could fly great distances thanks to the innovation of inflight refuelling systems. These planes were huge, they had swept-back wing designs to reduce drag (a design feature that had come directly from the drawing boards of designers working during the final years of the war) and they were built to last, to perform in raid after raid. When coupled with the power of the jet engine, these designs quickly evolved into the first jet airliners.

The British Overseas Airways Corporation (BOAC), now British Airways, was the first airline to put a jet into service. The De Havilland Comet, a direct descendant of the bomber and the jet fighters manufactured by Gloster and Heinkel during the 1940s, first flew passengers in 1952. It travelled at around 550 mph, it could transport passengers non-stop to any European destination and within a few years it was to become the first plane to fly the route between London and Johannesburg. Indeed, the Comet was set to become the standard airliner of its day when a series of disasters caused by metal fatigue grounded the plane and set the British aviation industry back by several years.

While De Havilland and BOAC were recovering, American aircraft manufacturers had succeeded in convincing the airlines that they could fly jet airliners at a lower cost than the propeller planes still in regular use. The first American commercial jet transporter was the Boeing 707, which Pan Am operated with great success. The 707 was a radical aeroplane that burned kerosene, had four gigantic engines each capable of producing 17,000 pounds of thrust and could transport 181 passengers across the country in relative comfort. From this aircraft evolved all other Boeing planes including the world's most popular airliner, the 747.

In a parallel development, another radically different form of aircraft saw military service during the late 1940s and quickly became an everyday civilian vehicle. The earliest working helicopter was built by a Spanish engineer named Juan Cierva who first flew what he called an autogiro in 1923. In its earliest form, Cierva's autogiro was not a very practical machine, primarily because it could not be made to hover in one spot and it could only carry a pilot short distances. Although Cierva's invention did much to inspire later designers, his machine was not taken very seriously (although it was revived during the 1980s and spawned the sport of autogiro racing).

German aviation engineers were at the forefront of helicopter design during the early days of the Second World War and their first military model was the Focke-Achgelis Fa 61, which could fly at up to 70 mph and was used in 1945 to assist troops stuck in remote regions during the German retreat across Europe.

The man who turned the helicopter from a novelty item into a mainstay of military and civilian avionics was the Russian-born Igor Sikorsky who emigrated to the United States in 1919 with the intention of building a radically new form of aircraft. As a child Sikorsky had made wooden models of helicopters and before arriving in America he had spent a decade studying the theory of how such machines could fly. Financed by the United States military, Sikorsky finally succeeded in building a full-size piloted rotorcraft, the VS-300, which made its maiden flight in 1939.

Mirroring the arrival of the jet, the helicopter came too late in the war to be of more than very limited military use. The first mass-produced helicopter, the R-4B, was not commissioned until late in 1944, but after it was accepted by the military it evolved rapidly, filling a niche that no other aircraft could occupy. From the early 1950s, the helicopter went through a series of rapid developments all financed by military planners who realised that it could be used to drop troops into remote areas, that it could land and take off from the tops of buildings or from a ship and that it could carry large cargoes. In the civilian world the helicopter became essential to emergency services and aid organisations working in isolated regions.

It might be argued that of all human inventions the aeroplane has, along with the computer and motor car, changed modern society more than any other machine. It is staggering to think that little more than one hundred years ago the idea that powered flight might be possible, let alone a common everyday occurrence, was beyond the imagining of most people. There are people alive today who were very young children when Wilbur and Orville Wright took off from a sand bank in North Carolina one chill December morning. Today those same people can traverse the globe in hours.

Since 1903 the world has shrunk dramatically thanks to the development of aircraft. Today, we can reach anywhere in the world within a day and the aviation industry continues to be one of the largest in the world, a part of all our lives. Almost all the important advances in the design of aircraft have come from times of war, most notably since 1939. But economics has also played a major role in shaping the way we use aircraft. It is interesting to note that with the exception of Concorde (no longer in service) airliners of today travel only a little faster than airliners of half a century ago. This is because the cost of making planes that can fly faster and also carry a large number of passengers is prohibitive. Decades ago, airlines discovered what the public wanted: aircraft that could transport them at a reasonable speed, in comfort and safely, but all at the lowest possible price. Since these requirements have been met there has been little interest in supersonic airliners.

It is possible that in the not-to-distant future we will see hypersonic airliners that emulate incredibly fast military jets such as the Lockheed SR-71 which has attained speeds up to 2200 mph, or even the new NASA scramjet, the X-43A, which recently flew at Mach 9.8 (7000 mph). But to achieve such velocities, new engines and fuels will have to be developed that can allow designers and engineers to build craft that not only travel extremely fast but can carry hundreds of passengers at a cost equivalent to the airliners of today. But, so far at least, the public have not demanded faster air transport.

The Rocketeers

The first military rocket was a bird. Ninth-century Chinese military strategists hit upon the idea of tying bamboo tubes filled with the newly discovered wonder substance later known as gunpowder to the backs of birds which were then trained to fly into the ranks of an advancing enemy. Of course, this technique often backfired as the birds were notoriously fickle in choosing targets for their kamikaze missions. At about this time the first fireworks appeared in China. They were used during festivals and public ceremonies. But surprisingly, given their penchant for creative ways of killing the enemy, it was many years before the Chinese thought of adapting the firework into an effective weapon.

Improbable tales have been handed down telling of the often bizarre applications of rockets. One of these stories describes how a wealthy nobleman named Wan-Hu ordered workmen to make him a chair with forty-seven rockets attached to it. After settling into his home-made craft, he instructed forty-seven servants to light the rockets simultaneously. When the smoke cleared, it is said, nothing remained of Wan-Hu.

In Part 2 we examined how the Chinese fashioned fire arrows which they propelled from barrels using gunpowder, and these were certainly in use by the end of the first millennium. The earliest description of what sounds like a self-propelled rocket rather than an arrow launched from a barrel comes from the Battle of Kai-fung-fu in 1232, in which an eyewitness describes '. . . thunder that shakes the heavens'. He further observed that a single device could scorch an area of over two thousand square feet. The Battle of Kai-fung-fu saw the Chinese ranked against a powerful Mongol army. In large part thanks to their rocket technology, the Chinese were victorious but the secret of the rocket was out and within a decade or two the Mongols had built their own devices and had begun to use them on the battlefield. By the end of the thirteenth century, knowledge of rocketry had flowed west, and as Arab traders and philosophers travelled through Europe they took the military ideas of the Chinese with them.

Among the most important Arab military texts were the writings of al-Hasan al-Rammah in which he described the Chinese and Mongol rockets. These devices were certainly in use in Italy by 1285, and, a century later, the word *rochetta* had entered the Italian lexicon. But just as the ancient Chinese before them, thanks largely to the fact that the rocket appeared on the scene at about the same time as the cannon (a weapon considered more versatile and easier to use), the Europeans of the Renaissance regarded it as little more than a toy.

Complementing the hugely successful gun, by the fifteenth and sixteenth centuries rockets were being employed in European battles. It is even said that Joan of Arc deployed missiles during her defence of Orléans in 1429. During the following centuries, many books dealing with the use of the rocket as a tactical weapon appeared. Yet it is striking that during this time many more texts were written illustrating the use of rockets in festivals and pageants. One of the most famous of these was *La Pyrotechnie*, written by Hanzelet Lorrain in 1630; this contained a certain amount of information concerning military applications, but most of it was filled with designs for rockets to be used on feast days and for other public celebrations.

Military men only really began to take the rocket seriously during the eighteenth century. The French were leaders in the field and

Napoleon is known to have been keen on their use. Most strikingly, rockets as weapons became very popular in India during the late eighteenth century and they were deployed with great effect against the British. It is likely that a tradition of rocket design for ceremonial use and public celebrations had taken root in India as the knowledge of the Chinese spread west during the thirteenth century, but it is clear that the Indians were quicker than the military minds of Europe in applying this technology to the development of weapons. During the long slow advance of the British into India during the final two decades of the eighteenth century, the redcoats learned some bitter lessons at the hands of Indian rocketeers.

At the Battle of Seringapatam in 1799, the commander of the Indian defenders, Tipu Sultan, took to the field with a force of six thousand rocketeers. Their weapons were quite primitive affairs, but their sheer number caused considerable damage as they were fired low to the ground straight at the ranks of the advancing British infantry.

British military strategists should have been more aware of this weapon in the hands of their enemy. Almost a decade before Seringapatam, the Scottish traveller Quentin Craufurd had described the effectiveness of the rocket in a widely read travel book published in 1790: 'It is certain that even in these parts of Hindostan never frequented by Mahommedans or Europeans,' he wrote, '. . . we have met with rockets, a weapon which the natives almost universally employ in war.'[1]

However, the British gained valuable experience from the battle and within a decade the army had a well-equipped rocket brigade that saw action in 1806 in Bologne and in Copenhagen the following year. During this latter campaign, some 25,000 missiles were fired at Copenhagen and caused widespread devastation and panic. One soldier who witnessed the event (and apparently had not heard of the raid on Bologne) reported: 'I rather think this was the first time Congreve rockets have been brought into play, and as they rushed through the air in the dark, they appeared like so many fiery serpents, creating, I should think, terrible dismay among the besieged.'[2]

The rockets used on these occasions had been designed by Colonel William Congreve, an officer in the British Army stationed at the Woolwich Arsenal. His model was based upon captured Indian rockets. It was a simple device, the most basic rocket possible in fact, merely a tube filled with gunpowder and mounted on a stake with a fabric fuse protruding from the main body of the device. Congreve's design was quickly adopted by the military establishment and came in a range of sizes from the smallest, the eighteen pounder, to the largest weighing in at three hundred pounds.

Until it was superseded by the Hale rocket in the middle of the nineteenth century, Congreve's design was the mainstay of many rocket brigades that were springing up around the world during the Napoleonic era. They were used by the British Army to attack Callao in 1809, Cadiz in 1810 and Leipzig three years later. Most famously, they were deployed during the War of 1812 and were immortalised by Francis Scott Key in what became adopted as the American national anthem, 'The Star-Spangled Banner': 'And the rockets' red glare, the bombs bursting the air, Gave proof thro' the night that our flag was still there', written to commemorate a battle between the Royal Navy and the American defenders of Fort McHenry. However, one British commander on the scene, Lt. George R. Gleig, saw the occasion from an entirely different perspective, remarking that he 'never did see men with arms in their hands make better use of their legs'.

Soon after the earliest use of the Congreve rocket in war, it produced spin-offs within civilian life. Whalers began to replace their guns with Congreve rocket launchers and another Englishman, Henry Trengrouse, demonstrated how they could be used to help endangered seamen. In late December 1807 he had witnessed a great sea disaster when the Royal Navy frigate HMS *Anson* ran on to rocks in full view of those watching in horror from the beach only three hundred yards away. One hundred and ninety members of the ship's crew of three hundred died that night.

Trengrouse concluded that a rocket could be used to fire a lightweight cable to a ship in distress. The cable could then be tied to a rope to form a lifeline from ship to shore. Very soon after this rockets began to be used as flares and warning signals and were only relegated to last-resort emergency use with the advent of radio and Morse code.

Within a few decades of rockets finding peaceful uses and saving lives rather than taking them, they began to fall out of favour with the military. It is easy to see why. Guns were becoming lighter, cheaper and more accurate. The arrival of mass production meant that literally millions of identical firearms could be churned out; they were safer to use and required less training to fire them effectively; cannon and side arms were becoming universal weapons.

During the First World War the British, American, French and German armed forces all used rockets but only in a very limited capacity. By this time weapons designers were attempting to develop a multistage rocket that could travel several miles and deliver payloads larger than any conventional cannon, but the technical difficulties of achieving this were formidable. When the fighting was over, there was understandably little desire to expend energy on new armaments. As a consequence, the rocket was a baton handed to civilian engineers and scientists who believed that it offered an exciting and inspiring future. These were people who perceived the rocket as nothing less than a way to transport payloads beyond the earth, a device that could eventually enable man to reach the stars.

During the last decades of the nineteenth and into the early twentieth century, science fiction boomed. Jules Verne and H.G. Wells both wrote a string of popular novels that not only sold well but were extremely influential. Most of the science described in such works as Wells's *The First Men in the Moon* (1901) and Verne's *A Trip to the Moon* (1902) was rather inaccurate, but these books served as a springboard for imaginative scientific minds all over the world, spurring on young men who were thinking seriously about space travel.

In Russia, a thirty-one-year-old, small-town teacher and science fiction enthusiast named Konstantin Tsiolkovsky wrote a ground-breaking article in 1898 called 'The Investigation of Outer Space by Means of Reaction Apparatus'. When it was finally published in 1903 in a magazine entitled *Science Survey* it went largely unnoticed but almost everyone who did read it dismissed its contents as pure fantasy. In this article, which was later expanded into a book, Tsiolkovsky described the basic requirements for a spacefaring rocket, the kind of suits astronauts would wear, the means by which the pilots of a rocket could breathe, as well as how they could move around and be supplied with food and water. Tsiolkovsky's was a futuristic vision, decades ahead of most of his contemporaries. He wrote:

> To place one's feet on the soil of asteroids, to lift a stone from the moon with your hand, to construct moving stations in ether space, to organise inhabited rings around Earth, moon and sun, to observe Mars at a distance of several tens of miles, to descend to its satellites or even to its own surface – what could be more insane! However, only at such a time when reactive devices are applied, will a great era begin in astronomy: the era of more intensive study of the heavens.[3]

Elsewhere he described how much he owed to the inspirational quality of Jules Verne's ideas. 'For a long time, like everyone else, I viewed the rocket from the standpoint of amusements and small applications,' he wrote in 1911. 'I do not remember exactly when the idea came to me to do calculations relative to the rocket. It seemed to me that the first seeds of the idea were cast by the famous fantasy writer Jules Verne; he awakened my mind in this direction. Then desires arose and they were followed by the activities of the mind, which of course would have led to nothing had they not encountered the aid of science.'[4]

Tsiolkovsky was a theorist, and a very good one, but he was never involved with any plan to build a rocket, nor did he conduct any genuine experiments. This duty fell to a very different character, an American named Robert Goddard, a man with grease on his hands and a mind that could encompass both intricate theory and sound practical knowledge.

If one individual may be said to have been the earliest progenitor of a practical long-range rocket, a model for the spacecraft of later generations, it is Goddard. Another genius years ahead of his time, he was responsible for many of the firsts in rocketry, and during his career he was granted more than seventy patents linked with rockets and space travel.

Goddard obtained a Ph.D. in physics in 1911, and while teaching physics at Clark University in Worcester, Massachusetts, he began to develop many of the mathematical models for spacecraft design and the mechanics of travelling beyond the earth. By the early 1920s, with funding from his college as well as substantial financial support from the Smithsonian Institution, he had begun building rockets and investigating the possibilities of liquid-fuelled vehicles. These would later supply the template for the V1 and V2 as well as the Saturn V launcher of Apollo and all rockets used until the present day.

At this stage the general public had only really been introduced to the idea of space travel in science fiction and fantasy novels, but it was Goddard who unintentionally awakened a slumbering fascination for the notion and drew to him both positive interest and the scathing ridicule of many doubters. In 1920, he compiled a lengthy and detailed scientific report on the possibilities of his research in order to acquire a grant of $5000 from the Smithsonian. Most of this paper was focused on the purely analytical scientific aspects of his researches to date, along with his hopes for the future, but towards the end of the proposal he described how his rocket could, given suitable funding for a great deal of further research, take a man to the moon.

Of the copies made of Goddard's proposal, a few reached Europe and Asia, and within months he was receiving letters from those enthusiasts who wanted to go to the moon at the earliest opportunity. At the same time, newspaper articles appeared criticising and ridiculing Goddard's ideas and reporters and editors across America, keen to boost their papers' sales, called upon any expert they could find to refute his ideas. Goddard's response was pragmatic. 'All I want to do is get this thing off the ground,' he was quoted as telling one sceptical journalist.[5]

One of the most remarkable of the critical comments on Goddard's ideas came from an editorial in the *New York Times* of 13 January 1920:

> As a method of sending a missile to the higher, and even to the highest parts of the earth's atmosphere, Professor Goddard's rocket is a practicable and therefore promising device. It is when one considers the multiple-charge (multi-stage) rocket as a traveller to the moon that one begins to doubt . . . for after the rocket quits our air and really starts on its journey, its flight would be neither accelerated nor maintained by the explosion of the charges it then might have left. Professor Goddard with his 'chair' in Clark College and countenancing of the Smithsonian Institution, does not know the relation of action and reaction, and of the need to have something better than a vacuum against which to react . . . Of course he only seems to lack the knowledge ladled out daily in high schools.

But it was the editor of the *New York Times* who was completely misinformed, not Goddard. He was mistakenly under the impression that a rocket flies because the exhaust gases 'push' against the medium through which it is travelling, and that if this medium was to become a vacuum (by the rocket flying too high) it would have nothing to 'push' against and would therefore be, as he put it, 'impracticable'.

But that is not how a rocket works. The best way to visualise how a rocket does fly is to think of a child's balloon. The air inside a balloon consists of particles moving at over three hundred metres per second, yet while it is sealed up the balloon stays in one position. This is because the particles of air in the balloon are bombarding its inside surface uniformly. If you release the air from the balloon it will, of course, shoot rapidly and randomly around the room. This is because the air inside the balloon is no longer pushing against the inner surface evenly; in one part of the balloon the air is rushing out, so the air pushing in the opposite direction will move the balloon that way.

A rocket may be considered in the same way. If it was a sealed cylinder it would not move even though the gases in the reaction chamber bombard the walls of the rocket; it would simply explode. But if the heated gases inside the rocket are allowed to rush out of a nozzle at the base, the gases pushing against the inner surface of the rocket in the opposite direction propel it skywards.

Its motion is caused by entirely self-contained factors; it needs no outside medium to 'push against'. Although misinformed criticism from the likes of the editor of the *New York Times* was understandably galling for the serious minded Goddard, the attention of the media served to increase the military's interest in rocketry, one that had waned considerably during the first decades of the twentieth century. Goddard's inventive work did not reach the demonstration stage until 1926 (and even then his earliest rockets were abject failures), so he was too late to influence military planners during the First World War. Then, during the immediate postwar years, interest in military matters died down and Goddard's work was focused on purely scientific applications (hence his need for funding from such organisations as his own college and the Smithsonian).*

However, even as early as 1916, long before Goddard had launched his first successful rocket, he had been forward-thinking enough to grasp two important facts. First, he would only be able to complete his experiments if he was given substantial funding,

---

*There is an amusing footnote to this episode. On 17 July 1969, as Apollo 11 was on its way to the moon, the *New York Times* published an addendum to its editorial of 13 January 1920 in which it admitted: 'Further investigation and experimentation have confirmed the findings of Isaac Newton in the 17th century, and it is now definitely established that a rocket can function in a vacuum as well as in an atmosphere. *The Times* regrets the error.'

whether from academic institutions or from the military; and second, he should put his stamp upon his creation from the very beginning if his contribution was to be recognised. Both of these facts are clear from a letter Goddard wrote to a friend in 1916: 'The finished article should be referred to as the "Goddard Rocket", or something of that sort,' he began while describing the hoped-for rewards from his efforts. 'This is because, as I told you, I wish eventually to use the thing for scientific work, and inasmuch as this promises to be expensive, I shall probably have to call for grants or subscriptions.'[6]

Goddard was only employed by the American military in the immediate lead up to America's entry into Second World War and even then they did not make the most of his talents. Seen as too eccentric, too left field to be useful in the present war, many of Goddard's suggestions were ignored or passed over. From 1942 until his death three years later (a few days after the atomic bomb was dropped on Hiroshima), Goddard worked for the US Navy assigned to designing a rocket-powered fighter plane, the JATO (jet-assisted take-off), which ended up in a military engineering cul-de-sac.

Because of this misplacing of Goddard's talent the rocket was not designed by the Allies in conjunction with the Manhattan Project. Instead, a combination of scientific brilliance and the willingness of a government to invest heavily in rocket research for the purposes of war enabled Germany to make the greatest breakthroughs in rocketry and lay the foundations for the technology that created the Space Age, and, later, the huge success of NASA and the Soviet space programme (which also employed captured German scientists).

The scientist who really made possible the American space programme was Wernher von Braun, another young man who had been inspired by the fantasies of Jules Verne and H. G. Wells and who dreamed of travelling to the stars, building space stations in orbit and constructing manned bases on the surface of the moon. But instead of forging reality directly from these dreams he became the head of a Nazi project to build missiles with which to attack

London, Antwerp and other Allied cities, and which would, if the war had lasted a few more years, have been capable of reaching New York. Only later, with the knowledge he acquired building weapons paid for by Hitler's government, could von Braun turn his skills and researches to peaceful solutions and to design and help build the first rocket to take men to the moon.

Von Braun began his career as an enthusiast but he was also a brilliant academic who combined mathematical ability with a gift for engineering and a clear ambition. Since boyhood he had been steeped in the ideas of rocketry and space travel. At the age of thirteen he read *The Rocket into Planetary Space*, a book written by fellow German enthusiast Hermann Oberth, then a leading light in the Society for Space Travel (Verein für Raumschiffahrt or VfR), formed from a group of keen Berlin amateurs who raised money to finance the manufacture of their home-made rockets. Five years later, in 1930, von Braun joined the VfR and he very quickly became a key member of the group, working directly with Oberth to design and build bigger and better rockets and launching them from a patch of land owned by the society in Berlin.

Thanks to the rise of the Nazis, things moved fast for the rocket enthusiasts during the early 1930s, and the interest of one particular high-ranking official named Walter Dornberger, who was then a high flyer in the Nazi Party, helped bring the society to the attention of influential military strategists. Dornberger was particularly drawn to von Braun, whom he believed had the makings of a great scientist, and it was in no small part due to him that by 1934 the twenty-two-year-old had been made an officer in the German Army, acquired a Ph.D. in aerospace engineering and was the key player in the Nazis' plans to build an advanced rocket weapon.

Hitler approved funding for the project, and in 1936 the army established a dedicated site at Peenemünde on the Baltic to design, test and launch rockets. It was from here that von Braun took some of the early designs dreamed up by the amateurs of the VfR and transformed them into workable devices that led eventually to the

V2 missiles, launched in their hundreds between September 1944 and May 1945.*

Von Braun was never entirely trusted by the German military. He was imprisoned for a short time, accused of spying for the Allies, and saved from execution only by the interference of his mentor, Dornberger. Although there have been claims that von Braun was a Nazi himself and a Hitler sympathiser, there is no hard evidence to prove this and it is revealing that he and his team were considered so dangerous that, as the Reich crumbled, orders were given that the team responsible for the V2 should be exterminated rather than be allowed to fall into enemy hands.

Von Braun became aware of this order (possibly thanks again to Dornberger) and as the British, Americans and Russians descended on Germany in the spring of 1945 he and most of his team and their families went into hiding in the small town of Bleicherode in the Harz Mountains. It was here in May 1945 that he surrendered to advancing American troops. From there, the rocket team was debriefed and interrogated before being transported to London and then on to the United States.

At this moment in history an odd series of events began to shape man's space adventures for at least the next two generations. Most of the scientists in the Peenemünde team travelled with von Braun to Bleicherode, and from there to the United States where they established new careers and homes, but a few who remained at Peenemünde were captured by Russian troops. More importantly, when von Braun and his team left their rocket base they could take with them only a portion of the vast collection of files, plans and other documentation relating to the V1 and V2 projects.

---

*Although the V1s and V2s developed by von Braun's team terrified the British public at a point in the war (immediately after the D-Day landings and the liberation of Paris in the summer of 1944) when it was thought that things were going especially well for the Allies, they had a minimal material impact on the war. Nine thousand two hundred and fifty-one V1 rockets were launched on Britain during 1944 and 1945 (of which 4261 were destroyed by the RAF). Those that got through killed almost six thousand civilians. One thousand one hundred and fifteen V2s were launched during the final months of the war, resulting in almost three thousand civilian deaths.

Consequently, a substantial amount of data concerning the building of rocket weapons fell into Russian hands. As the Cold War began, the Americans had the majority of the German team working for them but the Russians had a large archive of information with which they could build their own weapons and develop their own space programme.

The race began almost immediately. In America, the Peenemünde scientists made rapid progress thanks to almost unlimited resources provided by the military. Their first base was at White Sands testing ground, linked to Fort Bliss in El Paso, Texas, but in 1949 the group, along with hundreds of ancillary staff, moved to a facility at the Redstone Arsenal at Huntsville, Alabama. Here, the team (by now all American citizens) developed the early Redstone rockets that became some of the first launch vehicles in the space programme and the core of the ballistic missile arsenal of the United States. The Korean War of 1950–3 gave an additional boost to the enormously expensive development funding for these rockets.

In 1955, just a decade after the end of the Second World War, President Eisenhower pulled together the resources of the US Army and Navy to establish a joint forces project to build a medium-range missile that could travel 1500 miles. The result was Jupiter, the rocket that launched America's first satellite, Explorer I, on 31 January 1958. Jupiter was the direct predecessor of Saturn V, the rocket used to launch men to the moon little more than a decade later. This rocket also provided the backbone to Polaris, and it evolved into the launch system for its successor, the Trident ballistic missile.

Ironically, however, in spite of the fact that they began with far fewer of the original Peenemünde team and with the data on file rather than in the heads of the world's top rocket designers, for at least two decades after the end of the Second World War the Russians were streets ahead of the Americans in both the space race and the intercontinental ballistic missiles race. This was partly because Russia had been designing rockets long before the data from Germany became available. Another related reason is that, from the outset, the Russians believed in the concept of the intercontinental rocket. In America, the influence of politicians who doubted the ability of the German émigrés to build a missile that could travel thousands of miles acted as a powerful hindrance. For a while, their attitude overshadowed the enthusiasm of those who wanted to develop a space programme as well as those who desired a new and powerful weapon that could be married to the other great military acquisition of the Second World War, the atomic bomb. In Russia, there were few such doubters.

The chief architect of the Soviet space and ballistic missile programmes was a man named Sergei Korolev. Although he was born in 1906, six years before von Braun, their lives followed parallel courses in that they both became heroes of their nations' space and military projects. The two men never met but Korolev was at Peenemünde immediately after the war and he helped transport back to Russia many of the important documents on rocketry. Throughout the 1950s and 1960s, Korolev, known throughout his career simply as the Chief Designer, was in sole scientific control of the Soviet space and military missile programmes.

During this time, the great Soviet space firsts, Sputnik 1 in 1957, the first probe to reach another heavenly body, Luna 2, which crash-landed on the moon in 1959, and Yuri Gagarin's first manned flight in 1961, put the Soviet Union far ahead of the Americans in launching both spacecraft and intercontinental ballistic missiles. Thanks to the Manhattan Project, the West was many years ahead in the development of atomic devices but the Soviets had better delivery systems and more warheads than the West. This was a frightening prospect for an American administration determined to maintain the image of a technologically superior West.

For many years both the Russians and the Americans kept their space research and weapons development programmes intimately entwined. This close relationship continues even today in Russia but in the United States Congress responded to the launch of Sputnik 1 in October 1957 by passing the National Aeronautics and Space Act. Later that year, the National Advisory Committee for

Aeronautics (NACA), founded in 1917, was transformed into the National Aeronautics and Space Administration (NASA) and given an official remit to '. . . provide for research into problems of flight within and outside the Earth's atmosphere, and for other purposes'. The words '. . . and for other purposes' are important in this statement because they define clearly what NASA did in those days and still does from time to time. It was established as a civilian organisation and given a role primarily of scientific research, but it had (and still has) intimate links with the military. All NASA pilots have had military training and most of its astronauts are military officers.

In the early days almost every American space launch placed a military payload into orbit, and this is an important part of the reason why America won the space race and continues to be the leading force in space exploration. The vast resources ploughed into NASA to help the military (primarily to provide them with surveillance and communication facilities) also gave a huge boost to the civilian space programme. Putting aside for a moment the broad recognition that the successful moon landing of July 1969 was spurred on by the propaganda front of the Cold War, the knowledge the US military accrued during decades of missile research was invaluable to the parallel development of civilian space technology.

In Russia the link between military and civilian space research was so strong that these two aspects of the same game were almost indistinguishable. In the beginning this close relationship was extremely advantageous to Russian science as money flowed into research from the outset. In the United States military and scientific work were less closely linked; as a consequence substantial funding for purely scientific space research only came after the government had spent billions on weapons development. Later though, the Russian system became less successful. With the sole exception of the effort put in to reaching the moon first, almost all Russian research was driven by direct military need, and the demands of the chiefs of the armed forces came first. Also, Russian research was conducted with obsessive secrecy, which was a serious limitation. By contrast, when the American space programme (and with it the space race with the Soviets) really took off, NASA drew in some of the finest minds from a great range of scientific disciplines and industry. When it came to their weapons development programmes, the American military were every bit as secretive as the Russians but their approach to their civilian space programme was far more open.

In May 1961, President John F. Kennedy took the first official and public step towards reversing the trend of Soviet superiority in space and missile research when he announced to Congress:

Recognising the head start obtained by the Soviets with their large rocket engines, which give them months of lead time, and recognising the likelihood that they will exploit this lead for some time to come, in still more impressive successes, we nevertheless are required to make some efforts of our own . . . I believe this nation should commit itself to achieving the goal, before this decade is out, of landing a man on the Moon and returning him safely to earth.[7]

Kennedy's speech immediately galvanised the American public, government and industry. According to Kennedy's scientific advisers, reaching the moon was an achievable goal, and although no one expected the project to be cheap, and no thought was given to the spin-offs that would come from the project, it was deemed to be the only response America could make to its universal enemy.

There are few better examples of the central thesis of this book than the space race set in motion by the Kennedy administration in May 1961. By the beginning of the twentieth century (a century which witnessed the most horrific wars known to humanity), rocketry had long since faded from the minds of most military thinkers. At that time, the rocket was perceived as little more than a toy or as fodder for science fiction writers and fantasists. But then, towards the end of the Second World War, the rocket was transformed into a weapon that offered the potential to change the military and political map of the world because it had been developed at precisely the same time as the atomic bomb.

The military took the rocket and turned it into a serious proposition. Even though they are expensive, their strategic and tactical importance are obvious. With remarkable speed, rockets were built to fly further, higher and faster than ever before and they offered a means of attacking a distant enemy without risking human lives. But at the same time this military interest promoted scientific research, and as a consequence society has benefited enormously.

The list of beneficial spin-offs from the space programme reads like a wish list drawn up by an extraordinarily far-sighted and imaginative science fiction enthusiast of the 1960s. This is partly because of the breadth of innovation required to launch a space programme, which results in a diverse range of technological paybacks.

The most obvious and perhaps most lucrative benefit of the space programme has been the satellite. Some 4500 have been launched since the flight of Sputnik 1 in October 1957. Seven countries other than the USA and Russia (Japan, China, France, India, Israel, Australia and the UK), have put satellites into orbit and international organisations such as the European Space Agency (ESA) have supervised independently the launch of hundreds of satellites. At any one time there are probably close to one thousand working satellites in orbit and many of these are military devices used regularly for spying, to facilitate reconnaissance and communications and to aid targeting of enemy positions. During recent wars, most particularly the first Gulf War in 1991–2 and the second in 2003, satellites played an integral part in the military programme of the coalition forces. They were used to verify targets, to enable military communications and to determine the success of an operation. Satellites also made it possible for the media to transmit live pictures from the war to hundreds of millions of homes around the world.

According to research conducted in 2001, the US space industry is worth an estimated $61.3 billion annually.[8] This puts it into the same category as many other top-ranking industries such as textiles, computers and agriculture. Only about 15 per cent of this valuation was accounted for in the cost of rockets, propellants and hardware for space travel itself. Over half of the total came indirectly from earnings linked with satellite use. Most noteworthy amongst these is the communications industry, which incorporates the vast majority of domestic and international phone, mobile phone and internet traffic. The industry generating this vast annual income supports half a million jobs, 90 per cent of which (450,000) are in satellite manufacture and services, making it one of the largest employers in the world.

All of this began with the competition between the United States and the Soviet Union to reach the moon first and to outdo each other in the development of intercontinental ballistic missiles. But the satellite and all it offers is just one aspect of the commercial and domestic gains from the Cold War.

From the start of the race in the late 1950s until soon after the last Apollo mission, that of Apollo 17 in December 1972, the media were keen to write about the spin-offs derived from space exploration. They frequently cited famous examples such as Teflon (which, although invented in 1938, was used extensively by the designers of the Apollo hardware and from there quickly entered the commercial world) and digital watches, which were essential to astronauts but were made available to the public within a few years for as little as $10 each. Yet, not long after the moon race had been won by America and the spotlight on space had been switched off, interest waned and reports about all the beneficial technology derived from the space race no longer occupied column inches in daily newspapers.

Medical science has been one of the greatest beneficiaries of the space programme. The first artificial heart was developed from technology used to design and build fuel pumps for the space shuttle. NASA technology also contributed significantly to improving CAT (computer-aided tomography) scanners and MRI (magnetic resonance imaging) technology, both indispensable medical tools.

The laser is a perfect example of a technology that was only in embryonic form when it was adopted by space engineers but which evolved quickly into one of the most versatile and now widely used devices of our time. Invented during the late 1950s at Bell Laboratories in Murray Hill, New Jersey and patented in 1958 by Charles Townes and Arthur Schawlow, the laser was quickly adopted by the military as a key to missile guidance systems and as a weapon in itself.* Later, the laser was further developed by NASA scientists, at the Jet Propulsion Laboratory at Pasadena, whose primary goal was to be able to communicate over inter-planetary distances. From their work came communications devices used recently on manned missions and as part of the equipment in unmanned probes that transmit back images from distant parts of the solar system. We use just this technology when we make a call on our mobile phone or log on to a search engine.

Solar energy is perceived as a potential solution to the pending energy crisis. It may not satisfy the expected energy needs of the future, but it could be used in the home and in industry in con-junction with a number of other 'clean' energy sources. The technology involved in turning solar energy into a useable form to power machines and to heat homes came directly from the space programme. First used to power satellites in the 1960s, once any vehicle reaches space, solar energy is the most important energy source available for use on missions not too far from earth.† Solar power provides energy for the electrical systems on board the space shuttle and it keeps the International Space Station func-tioning. One day this power source will be key to a manned Mars mission; on earth, meanwhile, solar power helps to provide energy to Third World villages and to power remote weather stations. It finds use in a vast range of applications where conventional energy sources are impractical.

---

*Much controversy surrounds the invention of the laser and who got to it first. Towne and Schawlow were the first to publish their findings in 'Infrared and Optical Masers', *Physical Review*, vol. 112, No. 6, pp 1940–9, December 1958, but in 1964 the Nobel Prize for Physics was awarded to three scientists for their fundamental research into the development of the maser-laser – Townes and two Russians, Nikolai Basov and Alexander Prokhorov.

†For a deep-space mission such as the Cassini-Huygens mission to Saturn, spacecraft do better with nuclear power generators, as they are too far from the sun for solar power to be effective.

In Part 7 I will consider one of the most important technological advances to have come from the space programme – the computer revolution. Such a development would have been unimaginable without the stimulus of space exploration. It is comparable in importance to the evolution of the satellite in that it has done much to reconfigure modern technology, communications and the global economy.

As stated earlier, one of the most profound aspects of space spin-offs is that they have come in a variety of forms and play a role in a wide spectrum of human activity. The satellite has revolutionised communication and helped the Third World (through weather forecasting, communications and resource detection, to name but three examples). The laser is used in a variety of everyday appliances from the DVD player to the laser scalpel and spin-offs applied to medicine have radically improved methods of treatment. But there are other less obvious but equally as numerous advances that have derived directly from space research.

Consider the advanced materials developed specifically for spacecraft construction; for example, the heat-resistant material used in the heat shield of the Apollo command module and for the nose section of the space shuttle. Other special products developed by space engineers include lightweight space suit fabric, engine materials and the linings of rocket engine exhaust nozzles. All of these so-called 'smart materials' were designed for particular purposes, but each has found application in the commercial world for building, clothing, medicine, cars, aircraft and electronics.

One of the most successful commercial aspects of this flow of benefits from space into other walks of life comes in the way space engineers are constantly being challenged by the need to redesign conventional machines and devices for use in space, where restricted manoeuvrability and weightlessness have to be taken into account. The remote control, the cordless appliance and miniature versions of orthodox machines have all derived from the space programme. In each case these innovations came about because of the need to adapt some traditional piece of machinery for use in an alien environment.

There is, though, one more thing that man has gained from a space race that had its roots in rivalry and the Cold War, one that has little to do with technology or labour-saving devices but which lies at the very heart of what it means to be civilised.

Carl Sagan once wrote: 'All civilisations become either space-faring or extinct.' He continued:

> In all the history of mankind there will be only one genera-tion which will be the first to explore the solar system, one generation for which, in childhood, the planets are distant and indistinct discs moving through the night, and for which in old age the planets are places, diverse worlds in the course of exploration. There will be a time in our future history when the solar system will be explored and inhabited by men who will be looking outward toward the first trip to the stars. To them and to all who come after us, the present moment will be a pivotal instant in the history of mankind.[9]

Although we can now see that Sagan's view was overly optimistic, the exploration of space is inevitable. And if we survive long enough it is inevitable that we shall colonise other worlds and reach for the stars. Just as inevitable was the fact that war drove man to the point where space exploration became possible. The development of the missile as a weapon of war gave the initial impetus to space exploration and a race between two super-powers locked in a Cold War ensued.

Interest in war and the need for nations to defend themselves launched man into space and they are still the driving forces today. Each year billions of dollars are spent on military research linked with rockets, satellites and embryonic space technologies such as the missile defensive shield that has grown out of the 'Star Wars' project of the 1980s. However, space exploration is too big and expensive even for the military. In future, the journey through the solar system and beyond will be funded only partly by military need. The rest of the money will come from industry, from commerce and from tourism.

These sources of funding will help propel man to the stars. It is right – indeed, it is, as Carl Sagan wrote, essential – that we develop a space-faring capacity, for if we do not we shall fail as a species. And if one day we can sit back and leisurely ponder the path that took us to distant stars, and if we can casually survey a multitude of worlds populated by human beings, it should not be forgotten that, as with so much of our technological history, we achieved such wonders thanks to the peaceful application of ideas and breakthroughs that had their sources in war and conflict.

PART 6

FROM

THE TRIREME

TO THE

OCEAN LINER

The oceans separate the continents and for many generations they separated men from fulfilling their aspirations. But once the art and science of shipbuilding had been mastered the world opened up as never before. This technological advance was largely fuelled by one community's perennial need to have better war machines than those of its neighbours, and this determination has led to the rise and fall of empires and a steady flow of submission and dominance.

The twentieth-century arms race between East and West, which was part propaganda and part brinkmanship, prompted an enormous number of technical innovations and it remoulded our society. In particular, it gave us the computer, the satellite and global communications. The naval arms races that have taken place down the centuries have improved weapons and influenced the course of wars and the fates of nations, but they have not altered society in the same way as the twentieth-century space and arms races; nor indeed can the naval contests mentioned in Part 6 be compared, in terms of their ultimate productivity, to the way in which the development of the gun, the aircraft or medicine has affected everyday technology. However, the evolution of the ship, the merchant navy and naval fleets have played a key role in shaping civilisation. War and conflict advanced the design of ships, the way they are powered and the roles they perform, and these developments have enabled man to explore every conceivable corner of the world, to develop trade, which in turn boosted the importance of the Industrial Revolution, to advance banking and to forge international relationships.

All the great cities of the world are built next to water. They all grew up around ports and harbours because, until the middle of the twentieth century, all trade was conducted using land or sea transportation, and all intercontinental trade, whether it was between Europe and Asia or Europe and America, relied exclusively on merchant shipping.

The oceans, seas, rivers and lakes of the world have provided the infrastructure for civilisation, for the two constants in human evolution are war and trade; these are inextricably linked and each has been facilitated by seafaring. Command of the seas gave nations huge power and the development of the ship has guided the ebb and flow of cultural, economic and military power, breathing life into the dreams of imperialists and turning tiny nations into global rulers.

Opening up the World

How long ago human beings first took to the water is impossible to determine with any accuracy. Most probably primitive peoples used first logs and then dugouts to traverse rivers in order to find better food sources or to attack their neighbours. As civilisation evolved, the innate urge to explore, to conquer and to settle drove the brave and the determined to find any means available to reach their goals, and so boats developed and they grew bigger, stronger and more efficient.

In all likelihood the first use for a boat was to carry out an act of aggression, so it should not come as a surprise that the water soon became another battleground. However, the usefulness of boats was strictly limited by the ability to build them large and sturdy enough for them to have any practical application. As we saw in Part 2, the availability of metals facilitated the manufacture of early swords and spears; in the same way, local supplies of wood made a huge impact upon a society's seafaring abilities.

A good illustration of this is the way the Phoenicians, who lived around the eastern end of the Mediterranean (now Israel and Lebanon), cut wood from their cedar forests to build large, fast and very strong ships propelled by scores of oarsman. Around 700 BC the Phoenicians learned how to fashion long planks that could be joined together and sealed properly. The seafaring ambitions of the ancient Egyptians before them had been limited by available supplies of wood so that their ships were always constructed from shorter planks, making them far less seaworthy and restricting their fleet to the Nile and its tributaries.*

The Egyptians did not build dedicated military vessels. Instead they used their merchant fleet for policing their territory. But, because they drew no distinction between military and trading vessels, any advances in ship design fed back into society immediately and were not exclusive to the military.

In volume II of *The History*, written during the fifth century BC, Herodotus described the Egyptians' interest in building ships:

Their boats with which they carry cargoes are made of the thorny acacia, of which the form is very like that of the Kyrenian lotos, and that which exudes from it is gum. From this tree they cut pieces of wood about two cubits in length and arrange them like bricks, fastening the boat together by running a great number of long bolts through the two-cubits pieces; and when they have thus fastened the boat together, they lay cross-pieces over the top, using no ribs for the sides; and within they caulk the seams with papyrus. They make one steering-oar for it, which is passed through the bottom of the boat; and they have a mast of acacia and sails of papyrus. These boats cannot sail up the river unless there be a very fresh wind blowing, but are towed from the shore: from this acacia tree they cut planks 3 feet long, which they put together like courses of brick, building up the hull as follows: they joined these three foot lengths together with long close set dowels; when they have built up the hull in this fashion they stretch crossbeams over them. They use no ribs, and they caulk the seams from the inside, using papyrus.

---

*Some historians claim that the Egyptians made extensive trips along the coast of Africa as early as 1800 BC. However, these must have been rare and brave ventures.

A good illustration of how the Egyptians applied military experience to improve their business interests comes from the way military engineers developed the sail. In their earliest sailing ships the Egyptians employed a fixed sail. This meant that during a military engagement involving hand-to-hand fighting the decks were cluttered and the sails caused an obstruction. An anonymous military engineer, probably sometime around 1000 BC, came up with the idea of rigging which allowed a ship's captain to raise and lower the sail and yardarm. This then led to the innovation of reef points (small lengths of rope that secure the sail to the mast) which initiated the development of the elaborate rigging found in later vessels.

The Greeks were keen sailors and, unlike the Egyptians, they designed ships that had distinct roles. Even though the crews of all merchant ships had to be armed and able to defend themselves against pirates and foreign navies, the Greeks built different vessels for trade and for war. Indeed, they were the first to explore the great potential of specially designed warships. They realised that large vessels could be used not only to attack and defend but also to transport soldiers to a battle site. During wars against the Persians and in quelling the Spartans, the Athenians were known to carry a large part of the army (up to six thousand soldiers) aboard a fleet of triremes.

Triremes were elegant and beautiful vessels and the Greeks appreciated the versatility of ships that could be propelled by sails, oars or both, depending upon the circumstances. However, even the Greek ships operating in the relatively calm waters of the Mediterranean were of little use far from the coast, partly because they were flimsy and difficult to manoeuvre and partly because of the dearth of reliable maps and accurate navigational aids.

Nevertheless, such trips were attempted and recorded. There appears to have been no shortage of young men who wanted to see the edge of the world, or to find new lands en route – perhaps some of these were the original Jason and the Argonauts. But they had to rely upon the stars, ancient, unreliable knowledge and hearsay, and they had only the ancient and unreliable technique of dead reckoning to guide them. For such voyages oared craft were of no use. A single oarsman can provide only one-eighth of a horsepower and the space taken up by the oarsmen reduced room for storing supplies, making such a voyage impractical. The other problem with oared craft is that they have to lie low in the water. This presents no great difficulty for ships working inland waterways or when travelling close to the shore, but in rough seas such ships are quite useless.

The first real naval exchange took place in 480 BC at the Battle of Salamis in September of that year. The Persians, led by their king, Xerxes, had won a succession of land battles the same year which had brought most of the peninsula under their control, but when Xerxes's fleet of some seven hundred vessels was trapped in narrow waters by a much smaller contingent of triremes, his navy was routed. Knowing that the only way he could maintain his new dominion was by constant supply from the sea, Xerxes withdrew to open water leaving the Greeks unscathed and in complete control of the eastern Mediterranean for many years to come.

The Romans maintained an enormous navy and they used a variety of specialised vessels to supply and help defend their empire. In its pomp, the Roman navy consisted of at least five hundred ships. For most of the time these vessels were devoted to transporting men and supplies to the Emperor's sprawling dominions, but in 31 BC they engaged in at least one major naval conflict, the Battle of Actium, which involved more than seven hundred vessels. The imperial victory that day halted the ambitions of Mark Antony and Cleopatra.

By the time the Roman Empire was at its peak, the traditional method of fighting at sea, which involved ramming, boarding and hand-to-hand combat on the decks of ships, was modified by the introduction of the *harpex*, or *harpago*, a harpoon that not only caused considerable damage and loss of life but enabled the attacker to fire grapples and to close alongside the enemy. This was the first time a projectile device was used on a ship and it played a significant role in the Roman victory at Actium.

However, with the exception of the *harpex*, there was very little progress in ship design, shipbuilding or naval weaponry during Roman times, probably because there was almost never any genuine threat to the empire from the sea. When Rome did fall, it was to aggressors attacking on land from the north, the Goths and Visigoths, who came from largely landlocked countries.

Surrounded as they were on all sides by seafaring rivals each trying to fill the power vacuum left by the fall of Rome, the military leaders of the Byzantine Empire of around AD 700 were far more innovative and resourceful than their Roman predecessors. There is no better illustration of this resourcefulness than their invention of what became known as Greek fire, a devastating weapon used primarily at sea.

Greek fire is thought to have been devised by a Syrian engineer named Callinicus sometime during the middle of the seventh century AD. A highly inflammable liquid was stored in reinforced pressurised barrels and propelled from a long, broad tube projecting from the side of the ship; the crew hid behind iron shields surrounding the siphon as it was fired at an enemy vessel. The formula for Greek fire was known only for about fifty years and then lost, but we can hazard a guess that it contained a very nasty blend of sulphur, quicklime, oil and magnesium. It was greatly feared by both those who deployed it and those on the receiving end, and it became notorious for the fact that the fire it produced (which could burn even under water thanks to the magnesium in the mixture) was extremely difficult to extinguish. Indeed, it is not difficult to visualise the dramatic power of this weapon, a power that must have seemed almost supernatural to the common sailor of the seventh century.

The first recorded use of Greek fire was when the Byzantine navy deployed it against an attacking Arabian fleet in their home waters in AD 672, and although no ancient description of the weapon in use has survived we may get a flavour of the terror it engendered from an account of the way a revived version of the original formula was employed on land during the thirteenth century. In his memoirs a French nobleman named Jean de Joinville, Seneschal of Champagne, observed the use of Greek fire during the seventh Crusade:

It happened one night, whilst we were keeping night-watch over the tortoise-towers, that they brought up against us an engine called a *perronel*, (which they had not done before) and filled the sling of the engine with Greek fire . . . When that good knight, Lord Walter of Cureil, who was with me, saw this, he spoke to us as follows: 'Sirs, we are in the greatest peril that we have ever yet been in. For, if they set fire to our turrets and shelters, we are lost and burnt; and if, again, we desert our defences which have been entrusted to us, we are disgraced; so none can deliver us from this peril save God alone. My opinion and advice therefore is: that every time they hurl the fire at us, we go down on our elbows and knees, and beseech Our Lord to save us from this danger.'

So, soon as they flung the first shot, we went down on our elbows and knees, as he had instructed us; and their first shot passed between the two turrets, and lodged just in front of us, where they had been raising the dam. Our firemen were all ready to put out the fire; and the Saracens, not being able to aim straight at them, on account of the two pent-house wings which the King had made, shot straight up into the clouds, so that the fire-darts fell right on top of them.

This was the fashion of the Greek fire: it came on as broad in front as a vinegar cask, and the tail of fire that trailed behind it was as big as a great spear; and it made such a noise as it came, that it sounded like the thunder of heaven. It looked like a dragon flying through the air. Such a bright light did it cast, that one could see all over the camp as though it were day, by reason of the great mass of fire, and the brilliance of the light that it shed. Thrice that night they hurled the Greek fire at us, and four times shot it from the tourniquet cross-bow.[1]

According to the great scholar of ancient Chinese culture Joseph Needham, the Chinese navy was using a version of Greek fire by at least AD 900. The formula for their weapon may have derived from the Byzantine Empire's, although it was quite possibly an independent and slightly different creation. This weapon, most historians agree, was superior to the earlier Mediterranean version because, more than a thousand years earlier, Chinese engineers had invented the double-acting piston bellows which could produce a continuous sheet of fire. This sophisticated device had also been used as early as the fourth century BC for spraying soldiers with clouds of poisonous gas.

Another factor in making the Chinese weapon more deadly and efficient was their superior knowledge of metallurgy which allowed them to make tough and durable siphons and storage containers out of the very best cartridge-quality brass, which contained 70 per cent copper. This meant that Chinese ships could carry more noxious blends with far less risk to their own crews.

In his book *Talks at Fisherman's Rock*, the Chinese historian Shih Hsu-Pai described how an equivalent of Greek fire was used in a naval battle on the Yangtze River in AD 975. 'Chu Ling-Pin as Admiral was attacked by the Sung emperor's forces in strength,' the author tells us.

Chu was in command of a large warship more than ten decks high, with flags flying and drums beating. The imperial ships were smaller but they came down the river attacking fiercely, and the arrows flew so fast that the ships under Admiral Chu were like porcupines. Chu hardly knew what to do. So he quickly projected petrol from flame-throwers to destroy the enemy. The Sung forces could not have withstood this, but all of a sudden a north wind sprang up and swept the smoke and flames over the sky towards his own ships and men. As many as 150,000 soldiers and sailors were caught in this and overwhelmed, whereupon Chu, being overcome with grief, flung himself into the flames and died.[2]

The Chinese were also very early innovators in the science of shipbuilding, first solely for building ships used to navigate inland waterways, but later as designers of merchant and military craft that could operate along the coast. No records of the earliest seafaring exploits of the Chinese have survived but certainly by the Sung Dynasty (AD 960–1279) the country had a vast navy numbering hundreds of ships and some 52,000 men.

Most Chinese naval vessels could not travel far from the coast because they were too flimsy to survive long in the often tempestuous South China Sea. Navigation aids were inadequate and inaccurate, and although by at least the twelfth century the Chinese had developed the science of map making, the oceans and distant lands remained largely uncharted. As a consequence, the Chinese Empire's sphere of influence was limited to the rivers and the ocean within a day's sail of the mainland.

Both Genghis Khan during the early thirteenth century and the Mongol leader Kublai Khan fifty years later built up the Chinese navy so that it became active in the dynastic wars that wracked China for centuries. Kublai Khan, undisputed ruler of the Mongol Empire (a dominion that stretched from Korea to Hungary and ruled over half the world's population), is said to have possessed a navy in excess of five thousand ships, and with his much sturdier vessels his vision was ever outwards towards new conquests.

Kublai Khan was an imperialist *par excellence* who understood the value of exploration as well as of conquest, trade as well as conflict. The naval force with which he attempted to invade Japan in 1281 consisted of some 4400 ships. He deployed a similar force to overwhelm Java a decade later. Much of China's great success as an imperial state during this era came about thanks to the rapid advances they made in ship design and the creation of better maps drawn from extensive surveys.

In the generations following Kublai Khan, the Sung were swept away by the Ming emperors but the navy maintained its important position, and, for a short time at least, the Chinese continued to look outwards to become keen explorers. They discovered India and Ceylon beyond the southern borders of their empire, and they established powerful trade links across south-east Asia and into the Pacific. China even had its own Columbus, a man named Cheng Ho, who, during the first half of the fifteenth century, charted the lands and the oceans west to India and north into Russian waters. Ho's ships were, on average, five times the size of those used by the Portuguese explorer Vasco da Gama (who mirrored many of Ho's endeavours but from the opposite direction some half a century later), and Ho could offer his emperor a vast fortune in claimed treasure as well as the new and lucrative trade routes he opened up.

But this extroversion was short-lived. In spite of all he had done for his country, by the 1430s Ho faced powerful enemies at the imperial court who succeeded in dissuading the Emperor from funding further exploration. At the same time, the Japanese, who had for centuries been military rivals, were building up their forces and they served to focus the attention of the Chinese ruler on matters nearer home. At the Ming court a battle ensued between those who wanted to press on with exploration, foreign conquest and trade and a stronger faction which demanded a more defensive policy. The victory of the latter group led China to turn inward, transforming it from a powerhouse of imperialism, trade and exploration to an isolated giant with no presence in global history for centuries.

Almost overnight, China withdrew from its outposts and ceased trading beyond its own waters, and the Emperor not only banned further exploration but prohibited the building of any seagoing ships. The tens of thousands of naval personnel who had crewed warships and trade vessels were transferred to ships and boats active only on inland waterways and the fleet was neglected. An indication of the power of this drive to retreat from the world may be gleaned from a speech written by the minister Fan Chi in 1426, two years before all Chinese ships were called home:

Arms are the instruments of evil which the sage does not use unless he must. The noble rulers and wise ministers of old did not dissipate the strength of the people by deeds of arms. This was a far-sighted policy ... Your minister hopes that Your Majesty would not indulge in military pursuits nor glorify the sending of expeditions to distant countries.

Abandon the barren lands abroad and give the people of China a respite so that they can devote themselves to husbandry and the schools. Thus there would be no wars and suffering on the frontier and no murmuring in the villages, the commanders would not seek fame and the soldiers would not sacrifice their lives abroad, the people from afar would voluntarily submit and distant lands would come into our fold, and our dynasty would last for 10,000 generations.

Fan Chi's point is understandable. He was probably sincere in his belief that this course would be the best for his people and he was certainly ignorant of the power of war to aid one's own society. His naivety, though, is revealed in the last sentence of this speech, where he suggests that foreign lands would automatically come to them without any prompting on their part. Of course, this did not happen.

In the year of this speech, 1426, China stood at a crossroads and comparisons with our own time and geopolitical situation, although not directly related, seem obvious. What to do? Look outward or inward? A global giant must tread carefully and heed the lessons of history. These questions are as relevant today as they were more than half a millennium ago, for although the details have changed, human nature and the human condition have not. After Chi's recommendations were embraced, China followed one of the possible routes that lay before it and embarked upon several centuries of decline. It disappeared from the world stage. This route, the very opposite of the one followed by China's Western counterparts, led to few military conflicts other than defensive ones against immediate neighbours, and so China's advanced proto-technological society was stopped dead, trapped like a fly in amber. While Spain, Portugal, Italy, France, Holland and England built and lost empires, industrialised their societies and reached outwards, China remained inward-looking, disinterested in conquest or fortune seeking, and it became so stagnant that during the twentieth and early twenty-first centuries it has been forced to struggle hard to regain centuries of lost ground.

And so, as China faded into obscurity, the countries of Europe, dragged from the Dark Ages and revitalised by knowledge from the East, entered a new era. They followed the other fork in the road ignored by the mandarins and Chinese emperors and along this route they found glory. The story of empire, the rise and fall of the southern Mediterranean states, the rise and fall of the Dutch Empire, the creation of a French power bloc and the global domination of Great Britain is a tale of military growth and the application of science. More specifically, in each case, it is a story of naval power and the application and use of that power to mould cultures, to spread influence, to build what may be called empires of reason, built upon the foundations of science and discovery.

Although many of the secrets of navigation, map making and ship design had been learned and developed by the Chinese, when they passed the baton of technological and social advance to the Europeans (via the Arabs) these innovations were reworked and revised and began to play a major role in the rapid and unstoppable emergence of Western culture.

Catalysts of Empire

Europeans had always been interested in seafaring. The Greeks and Romans had favoured oared ships called galleys, which used slave power. But in the states of northern Europe and Scandinavia, where the seas and oceans are more dangerous than the Mediterranean, advances were accelerated. The Vikings of the eighth and ninth centuries were skilled mariners who travelled long distances in vessels called longships that used both sails and oars. By the thirteenth century, their descendants, many of them experienced and well-travelled commanders, could take their crews as far as the Mediterranean, a vast distance for a sea journey at that time.

The Viking ships were much stronger and larger than anything the Greeks or Romans had built and they were an effective means of delivering armies to distant places. With very few rivals to offer competition within their sphere of influence, the Viking longships were only rarely used to fight at sea, and when they were they employed the ancient technique of ramming and boarding, tried and tested for centuries.

The countries of the Mediterranean caught up and surpassed the abilities of the Vikings and their descendants thanks to the influence of knowledge and experience garnered from the East. Much of this knowledge came from ancient texts written by classical scholars, engineers and marine adventurers. But in their desire to find original works of the classical tradition, Europeans extended themselves as far as possible towards the reported sites of wondrous collections and great intellectual treasure. Wealthy patrons such as Cosimo de' Medici in Florence financed these missions and did much to encourage the adventurous spirit of young men willing to embark on voyages into the unknown seeking intellectual reward, gold and glory. While these mariners were gathering information from the East, from Greece and from Rome, seeking the knowledge that had filtered down through the Arab writers and scribes who had kept alive an intellectual culture during Europe's Dark Ages, they also learned many things through hard experience.

For the sailor one of the most important innovations was the lateen, a triangular sail that was fitted to the mast with a moveable boom. This invention superseded the square sail previously favoured by all mariners, from ancient Egypt to Scandinavia. The lateen had been used in the Mediterranean since around the ninth century, its name deriving from the misconception that it had been invented by the Romans occupying Constantinople. But it had probably originated in China, another of the technological cues conveyed westward by the Arabs.

The lateen was one of the things that made possible the missions of the knowledge seekers of the thirteenth and fourteenth centuries, and at the same time it made oars increasingly superfluous. Because the boom was adjustable, in skilled hands ships with lateen sails could take advantage of changes in wind direction, providing both speed and navigational freedom.

Another essential development was the introduction of an advanced rudder. The stern oar, which had been employed since the time of the earliest boats, was a useful device for steering small ships, but as seafaring vessels grew larger, both for trade and for war, their heavy cargoes made it impractical. The modern rudder, or stern rudder as it was properly called, a device tucked low against the bottom of the ship and connected by pulleys to the bridge, is a further example of an ancient secret lost. First used by the Chinese of the Sung Dynasty, knowledge of its construction and use was passed on to the Arabs and introduced into Europe sometime during the middle of the thirteenth century. Evidence for this comes from seals used by German merchants from Weimar and Ebling.

At the same time as the rudder was making its mark, ships' rigging became far more sophisticated. The usefulness of the lateen had illustrated how a simple change to the structure of sails and the mast could greatly improve the speed and agility of a vessel. Gradually, all these refinements combined with the use of the more sophisticated rudder, transformed the ships of the time.

This transformation was one of the primary factors in the trade boom experienced in Europe between 1150 and 1300 when the price of most goods increased by 30 per cent.* But both the modern rudder and the lateen were also of enormous importance to naval forces, and by the thirteenth century a wide range of different ships from many nations could be found plying the waters of the Mediterranean. Oared vessels were still used but were becoming less common. Merchant vessels armed with crossbows and spears for defence against pirates could double up as naval vessels if necessary. Naval vessels had become a mainstay of the armed forces of all the countries with access to the sea, and ports began to grow in importance as they became centres of trade and military operations.

From about AD 1200 ships grew steadily larger, not only because they were easier to steer (and faster because of the lateen), but because the means of constructing them had improved greatly. The longships of the Nordic races had been built using large overlapping planks. This technique, known as clinking, was also employed by Mediterranean shipbuilders but because the wood available to these peoples was not as strong as Scandinavian pine

---

*It should be noted that although in the modern world we are used to inflation this phenomenon was uncommon during medieval times and a rise in prices of 30 per cent over a period of 150 years was highly significant.

and spruce the planks had to be cut shorter, thereby placing limitations on their design.

With increased trade, wood could be imported and ships of the Mediterranean changed accordingly. In turn, this meant that vessels could be built bigger and they could carry larger crews and more weaponry. By the 1400s, many merchant ships and naval vessels were several times larger than their counterparts from the thirteenth century. Double-masted ships became common and were able to carry more guns. Indeed, the trend towards bigger and more heavily armed ships continued until the twentieth century and became a hallmark of naval evolution.

By the fifteenth century, first the Spanish and Portuguese, and towards the end of the century, the English, had built up their navies to the point where they had become indispensable to national defence and strategic planning. The galleons of this period were magnificent constructions that dwarfed the vessels of a century earlier. These ships were designed for the sole purpose of fighting naval battles. At the same time, merchant ships had also grown in size and were carrying larger and larger cargoes in order to satisfy expanded commercial markets in foreign goods. These vessels needed protection from pirates (as well as rival navies) and so the different roles of the two types of ship that had once been intermeshed were again separated.

The navies of Europe were incredibly superior to those of Muslim forces, which had failed to move with the times. Ironically, the peoples who had acted as a conduit for so many technological advances (not just those linked with seafaring but over a far wider spectrum) were left behind during the fourteenth and fifteenth centuries. This opened the way for colonisation by the Spanish, Portuguese, English, French and Dutch, each of whom clamoured for global supremacy.

During the fifteenth century Spain and Portugal were the most powerful nations on earth. They ruled the seas and they became influential and very rich as the prime movers in international trade. At the same time, thanks to their associations with a global business network, the nation states of Italy also became wealthy.

Located as they were in the perfect geographical position to act as a hub between East and West, Venice and Genoa were the principal beneficiaries of this change in commerce. The success of the merchants and mariners of these and other Italian cities fuelled the Renaissance and left a legacy in the glorious art and cultural riches supported by stupendously rich patrons.*

But competition was fierce, and although Spain and Portugal had carved out empires based on trade and naval power, first the Dutch and then the English surpassed them. And in the case of the English (later the British after the Act of Union united England with Scotland in 1707) this led to the first globally influential empire since the fall of Rome.

Henry VIII was the monarch who began to shift the axis of naval power and to engender English dominance. Ascending the throne in 1509, Henry understood the importance of a strong navy for both defence of the realm and for the benefits of exploration and colonisation. His father, Henry VII, had no navy to speak of when he snatched the crown after defeating Richard III at the Battle of Bosworth in 1485, and, traditionally, English monarchs had employed mercenary sailors and hired vessels from other nations to fight their wars. It was a major shortcoming for any ambitious state or ruler. At the time of his death in 1509, Henry VII's navy consisted of just five warships. Henry VIII inherited these, and by the end of his reign in 1547 he had added thirty-five more to the English navy. Although small, this force (given the name the Royal Navy in the time of Charles II in the 1660s) was a formidable one thanks to the advanced designs and sheer size of the ships for which Henry had paid with money he had taken from the Church.

During this period, ships came with forecastles and aftercastles, and they positively bristled with weaponry. Some carried as many as 120 cannon, which were fired through the newly devised gun ports positioned low in the water to stabilise the vessel. Much about how to load a merchant vessel was learned from the need to

---

*Today 40 per cent of the world's art treasures are to be found in Italy.

pack a warship with guns, men and ammunition and some of these lessons were learned the hard way. The *Mary Rose*, Henry's most famous ship and his pride and joy, sank off the Isle of Wight in 1545 without a shot being fired during a skirmish with the French fleet whose commander was trying to land on the island just a few miles from Portsmouth. The ship was carrying seven hundred men (all of whom drowned) when it had a capacity of only four hundred, and it had been loaded with too many cannon and too much ammunition, so that when it was caught by a wave and started to take in water it sank in minutes.

Henry VIII was the founder of Britain's maritime tradition but it was his daughter Elizabeth and the great sailors and explorers with whom she associated, and whom she financed and encouraged, who took the navy from a worthy defensive force to one of global importance. Enmity between Protestant England and the Catholic states of Europe was a major feature of Elizabeth's reign, and the period was defined by great military and trading rivalry as well as vigorous competition between explorers seeking new lands to colonise.

Elizabeth's court could boast three of the most determined and successful mariners of the day in Walter Raleigh, Francis Drake and John Hawkins, each of whom played crucial roles in saving England from invasion, finding new lands to colonise and rewarding the support of their queen with vast riches stolen, discovered and murdered for. The New World, first discovered by Christopher Columbus half a century before Elizabeth came to the throne, was fought over by every leading nation in Europe: the Spanish and Portuguese, already in decline, the French with their expansionist dreams, and the Dutch and the English, then just starting their sharp ascendancy.

The turning point in this power struggle, and the single pivotal event that relegated Spain and Portugal to second-class status and elevated England to the condition of empire builder, was the defeat of the Spanish Armada in 1588. The English destroyed a much larger invasion force of Spanish ships thanks to a number of unconnected factors. The weather was with them and the planning of the Spanish was faulty, but the principal difference between the two navies and one that determined the outcome was technology. The English ships were better designed, more agile and faster, and under the guidance of the triumvirate of great English mariners the Queen's sailors had been better trained.

This technical skill and the application of more finely tuned shipbuilding abilities changed history. The English began to dominate the colonisation of newly discovered lands with competition only from the Dutch. Their influence moulded the shape of early America, the development of North Africa, Asia, and, later, India, Canada, Australia and the Middle East. Indeed, although the country where all this began has now returned to the position of a second-rate (albeit wealthy, stable and influential) power, the imprint of empire continues to provide the infrastructure for much of the world, from local government systems to the status of English as the most widely spoken language on earth.

Getting Around

The most important gain from the rise of the English navy and the power struggles between European states during the fifteenth and sixteenth centuries was the greater confidence and determination to explore, to seed and to expand horizons propelled by naval power. It was an age during which military prowess fed directly into the drive to colonise, to stretch beyond the confines of home. But it is important to remember that as well as the desire to explore, the pioneers and adventurers who crossed the seas needed practical guidance for they could never have succeeded in reaching distant shores if it had not been for important recent improvements in navigational techniques.

As late as the end of the fifteenth century, ships were still incredibly flimsy affairs and conditions on board were often intolerable. Ships were cramped, there was no real understanding of diet, so that most sailors were afflicted by dreadful illnesses such as scurvy (caused by a lack of vitamin C), and the need for adequate sanitation was not even considered. It was quite common for a significant proportion of a crew to die during the early voyages of discovery. They died from disease and from frequent accidents, they were murdered by pirates and natives, and many went insane. But although these men were determined, motivated by a lust for gold and often driven by ruthless captains, little could have been achieved without the invention of the compass and the map.

Both of these critical inventions had their antecedents in ancient times, but neither had been fully utilised and their potential had only ever been glimpsed. The first compass was a primitive affair, probably just a specially moulded spoon placed on a bronze plate. The name of this device, the diviner's board, indicates the original use of the compass. Linking the position to which the spoon turned under the influence of the earth's magnetic field to the details of the I Ching, the Chinese of the second century BC believed that the diviner's board could be used for prophesy.

In the eighth century book *Master Kuan's Geomantic Instructor*, a guide for the sage, the author tells us:

> The lodestone follows a maternal principle. The needle is struck out from the iron and the nature of mother and son is that each influences the other and they communicate together. The nature of the needle is to return to its original completeness. As its body is very light and straight, it must indicate straight lines. It responds to the *chhi* by orientation, being central to the earth and deviating in various directions. To the south it points to the Hsuan-Yuan constellation, hence to the Hsiu Hsing and therefore to the Hsiu Hsu in the north along the axis Ting-Kuei. The yearly differences follow the elliptic, and all such phenomena can be understood.

By this time, though, philosophers were already beginning to consider using the compass as a navigational tool and there is some evidence to show that, by the end of the first millennium, Chinese mariners were using a simple version of it to navigate. Later, the compass became essential to the success of the Chinese explorers of the early fifteenth century, most especially the afore-mentioned Cheng Ho, who travelled further and more widely than any of his countrymen before him. Its importance was recorded by both Cheng and his compatriot and fellow explorer Zheng He, who made seven international voyages during the first three decades of the fifteenth century.

News of the compass passed down the usual route of Arab interpreters and improvers during the Dark Ages until it was rediscovered in the West, probably around the thirteenth century. It is clear, though, that at this time few people understood how the compass actually worked. No distinction was made between true north and magnetic north and ships often ran aground or became completely lost even when they used the compass. It was only after the English scientist William Gilbert published his book *De Magnete* in 1600 that the properties of the compass and the difference between true north and magnetic north were understood.*

Up to this time, the compass was considered to be simply one of many rather careless ways of plotting a course at sea. For thousands of years sailors had used the position of the sun, the moon and the planets along with information drawn from star charts to guide them. Employing the cross-staff and the astrolabe, which had been used by the Roman philosopher and engineer Ptolemy in the first century AD, a competent seaman could calculate his latitude by reference to the North Star. But because ships do not tend to stay still but instead pitch and roll constantly even in relatively calm seas, these techniques were pretty crude; in stormy weather such means of navigation were totally useless.

The only other way of knowing one's position at sea was to use dead reckoning, which was only of any use if you stayed near markers and recognisable land masses. Another method was to head out using the stellar markers and to know how fast you were travelling so that you could calculate how far you had gone. The most common way to do this was to toss overboard a line with

---

*The magnetic north pole – the point you would reach if you travelled as far as you could go in the direction of north as indicated by a compass, is not the same as the true north. The magnetic north is the north pole of what may be thought of as a bar magnet inside the earth. It does not run precisely north–south along the true north–south line from pole to pole and it is also moving very slowly. Currently, it is to be found at about 82° of latitude and 114° degrees west.

knots tied at regular intervals. By timing how long it took for a known length of line to be laid, the speed of the ship could be calculated, hence the word 'knot'.

If it was difficult to calculate latitude, then the determination of a ship's longitude was almost impossible. By 1700, understanding of the compass had progressed enormously and the concept of true north versus magnetic north was appreciated to such a degree that charts showed what became known as magnetic variation (the degree of difference between true north and magnetic north) in many different parts of the world. These charts proved to be an essential aid for the mariner and all naval vessels carried them, making the compass an accurate and reliable navigational tool. On naval vessels the compass was housed in a sturdy box and protected as much as possible from the erratic movements of the ship.

Yet these improvements could only do so much. By the eighteenth century the Royal Navy dominated the seas and global British trade fuelled the expansion of the empire. But millions of pounds and many hundreds of lives were lost each year as ships went down at sea. Some vessels drifted into dangerous waters while others often took months to reach their destinations. This was mainly because it was almost impossible to determine accurately a ship's longitude.

This became such a problem that in 1714 the British government, through the Admiralty, established a prize of £20,000 for an invention that would enable a ship's captain to measure longitude to an accuracy of half a degree. To claim the prize the inventor had to '. . . pass over the ocean, from Great Britain to any such Port in the West Indies as those Commissioners choose . . . without losing their longitude beyond the limits before mentioned and should prove to be . . . tried and found practicable and useful at sea'.

It had been known since ancient times that for every 15° longitude one travelled east or west, the local time changed by one hour (because 24 × 15° equals 360°, or one complete circumference of the earth). So, if the difference between the time aboard ship and a predesignated standard (Greenwich Mean Time) could be established with great accuracy, a navigator could calculate the exact position of the ship due east or west. The weak point in this scheme was that the method required a timepiece of exceptional accuracy that could be carried on board a ship. In 1700, no such timepiece existed.

A body known as the Board of Longitude was duly established to administer and judge the Longitude Prize and it caught the imagination of the public to such an extent that 'finding the longitude' became a catchphrase of the day. Every inventor and crackpot, it seemed, expended huge efforts in attempting to solve the problem.* Even the ageing Isaac Newton tried to build an accurate enough clock but came nowhere close to succeeding.

One man who did succeed was John Harrison who spent the best part of his working life reaching the standard of accuracy demanded by the Board of Longitude. Harrison was a working-class joiner who inspired little respect from the scientific establishment, but he was immensely skilled, a determined perfectionist and he was driven by both the desire to create a 'perfect' clock and to cock a snook at the Establishment. In spite of numerous protests, and even the intervention of King George III, when Harrison eventually succeeded in constructing a timepiece that met the Board of Longitude's strict conditions he only received a part of the promised prize, and even this was not handed over until 1775, sixty-one years after the award was created and just twelve months before Harrison died, aged eighty-three.

Harrison was treated shoddily but at least he lived long enough to receive some financial reward, and he enjoyed the satisfaction of knowing he had achieved a great feat of design and engineering: it is no exaggeration to say his invention transformed navigation. Captain James Cook used one of his clocks on his voyage to Australia, and, within a decade of the Board of Longitude accepting Harrison's design, every British merchant and naval vessel was equipped with a fixed clock that could determine the ship's position by reference to Greenwich Mean

*A fact that is not at all surprising considering the prize money of £20,000 was a fortune in 1700, equivalent to several million pounds in today's money.

Time. Variations on Harrison's device remained the principal means of determining longitude until the twentieth century and the invention of radio.

However, knowing one's latitude and longitude is only one aspect of navigation. It is just as important to know where you are going, what to expect geographically when you reach your destination, and where you are in relation to surrounding land masses. In ancient times, such knowledge was usually passed on by word of mouth, something known by the Greeks as *periplous*. Sailors would offer such guidance as: 'Travel due west for three days until water of 70 fathoms is reached in fair grey sands. Then go north for one day and you will reach your destination.'

A few crude maps were available to the earliest mariners. These were guides that showed local landmarks and reference points. Surviving examples of these include Inuit maps drawn on skins and ancient Assyrian clay tablets dating from about 500 BC. In their very different ways these illustrate how ancient people represented their local area and how little they knew about lands beyond their borders.

The assimilation of maps in the West was rather slow. As with so many inventions and discoveries of the ancients, knowledge of map making was lost until the early Renaissance when centuries-old examples such as the first Greek map accredited to Anaximander and the Roman map of the world, the *Orbis terrarum*, commissioned by the Emperor Augustus in 12 BC, resurfaced. These were studied and the techniques employed in creating them copied.

The most renowned map-maker of the Renaissance was Gerhard Mercator who was born in Antwerp in 1512. By the 1550s, he had created detailed maps of Europe, Africa and Asia and had published several books on the subject of map making, including *Certaine Errors in Navigation*. These maps helped mariners enormously during an age before any accurate form of determining longitude was known, and ever since this time Mercator has been commemorated in the naming of ships, satellite navigation systems and even missiles.

Like many other advances in the technology of seafaring, the employment of maps by navies quickly led to their use by merchants and explorers, not least for the simple reason that many of those who served in the navy later became commanders of merchant vessels or set up business as explorers, taking their experience and knowledge with them. The British were particularly quick to see the importance of sea charts and maps. Immediately after the Jacobite Rising of 1745 they conducted the first topographic survey, the responsibility for which fell on the Board of Ordnance, which was the precursor to the Ministry of Defence. Today, producing updated digital topographic maps with the aid of satellites remains the job of the Ordnance Survey.

War has also propelled the advance of technologies that have made the making and interpreting of maps increasingly effective. One of the best examples of this is the development of aerial photography, which was a significant military development during the First World War, so much so that the newly invented aeroplane was used more as a means of photographing enemy positions than as a weapon of war. As both cameras and aircraft design improved, this technology assumed even greater importance, so that aerial photography played a key role in strategic planning during the Second World War.

Since the first space launches during the 1950s, one of the primary military roles of satellites has been reconnaissance. Reconnaissance satellites use specially designed cameras to photograph land masses in incredible detail, using not only optical cameras but also infrared photography and other advanced imaging techniques, all of which has enabled modern cartographers to produce maps of remarkable accuracy.

Satellites have also become a crucial element in modern methods of navigation. The GPS (global positioning system) was developed during the 1960s and 1970s by the US Department of Defense as a navigational aid for military aircraft and naval vessels around the world. The first system consisted of just seven navigational satellites in low polar orbit. By 1973, however, the Pentagon had allocated $8 billion to greatly improve the system, creating what was called NAVSTAR (navigation by satellite timing and ranging), which became fully operational in 1995.

By the late 1980s, GPS had become familiar to most people in the industrialised world, and it is now used in a wide range of civilian applications, from in-car navigation systems to air traffic control and the guidance systems on cruise ships. The earliest commercially available models were first sold in 1983 and cost $150,000; they needed two people to operate them. Today, hundreds of companies around the world make GPS equipment and a hand-held set costs less than $150, one-thousandth the original price.

Thanks to its enormous military importance and commercial versatility, both the military establishment and the civilian corporations that acquired access to GPS technology during the late 1970s are expending great energy in constantly refining the system, so that new and often startling applications are finding their way into our lives. The most advanced GPS systems can pinpoint an object or location anywhere on the surface of the earth to within about a metre. As this level of accuracy increases, the technology that guides cruise missiles and airliners is also beginning to be used by surveyors and designers, civil engineers and broadcasters. Linking GPS technology with the internet and mobile phone systems adds another layer of sophistication. One of the most interesting ideas for the near future is that of advertisers posting messages in 'thin air'. As we pass close to an invisible advertisement, the GPS network will alert our phone or wearable computer and send us an advertising message, a restaurant review or perhaps a quick movie preview. All of which offers a level of sophistication very far removed from the earliest uses of a compass to predict the future, or the first efforts to produce crude marks on a sealskin in order to create the earliest maps.

The Naval Arms Race

The foundations of the British Empire were laid in the six-teenth century but although it grew enormously over a period of three hundred years it could not really justify the term empire until about the beginning of the eighteenth century. Some histori-ans argue that the empire reached its peak when Queen Victoria was crowned Empress of India in 1876, but others claim that the power and extent of Britain's imperial rule reached its zenith around 1900. What was undeniably true was the maxim that the sun never set on the British Empire. Covering some fifty million square miles and holding dominion over an estimated five hun-dred million people (a quarter of the global population in 1900), in terms of land area it was the second largest empire in history after the Mongol Empire of the thirteenth and fourteenth centuries.

But although the British ruled supreme during the eighteenth and nineteenth centuries, they were not without their rivals. France experienced an industrial revolution and colonised large parts of the world almost contemporaneously with Britain, but they did not enjoy quite the same degree of success. The French Revolution of the last decade of the eighteenth century caused great internal upheaval and this disruption slowed their expan-sionism. From the ashes of the Revolution, Napoleon Bonaparte rose to prominence and he saw rivalry with Britain as the key to a greater future for the new republic. He built a powerful army and navy, and he tried, but ultimately failed, to achieve French superiority.

Among the weapons crucial to Napoleon were the navies (notably the Spanish fleet) his forces acquired through invasion and this combined force presented a serious threat to British naval domination. The Napoleonic Wars (1803–15) were waged on land and at sea, and although the conflict ended at the Battle of Waterloo in 1815, the struggles between the two nations reached a climax with Napoleon's plan to invade the British Isles in 1805 with a force of some 350,000 men. The Battle of Trafalgar in October that year not only scuppered his plans for invasion and sealed British naval dominance for a century, it was also the single most important naval confrontation of the nineteenth century. With victory at Trafalgar, Admiral Horatio Nelson joined the pantheon of British naval heroes, men like Raleigh, Drake and Hawkins, immortal seafarers and iconic defenders of the realm.

There were many reasons for the British success at Trafalgar. Credit must be given to Nelson whose abilities as a commander, tactician and mariner were superior to those of his French counterpart, Admiral Pierre de Villeneuve; but two other important factors contributed to the outcome. First, although the French ships were hardier than the British, the Royal Navy vessels were faster, more manoeuvrable, and, most importantly, they had superior fire power. Second, the British had the advantage of having developed short-barrelled, large-calibre guns called carronades (French spies had not yet discovered this valuable war secret). The carronade owed its development to the superior British industrial system. Although innovators, manufacturers and businessmen in other European states were quick to catch up with the originators of the Industrial Revolution in England, by the turn of the eighteenth century British industry and commerce were expanding with unmatchable speed. Two of the most significant elements of British industry were mining and iron production, which attracted many of the most energetic innovators of the time.

A third reason for the success of the Royal Navy at Trafalgar came down to lessons learned by naval administrators and military planners. In the late eighteenth century the Royal Navy was the only force in the world that worked closely with its government in almost every part of its organisation. The British government created a chain of communication with the Admiralty and offered credit and finance arrangements that enabled them to work together to supply not only the crew and the ships, but all the thousands of other items and commodities needed to keep a fleet functioning efficiently. This administrative sophistication had grown directly from the expansion of the British Empire in America and Canada where up to a hundred thousand men needed to be kept supplied with weapons, ammunition, clothing, food and many other necessities. To accomplish this, a team of military planners and civilian administrators had come together, thereby creating a system which kept the money flowing and the navy operating to its full potential.

The effort to sustain the colonies streamlined the running of the navy, but another important factor was its heritage. The Royal Navy had been created by Henry VIII and the strong ties between government and navy initiated during the sixteenth century had been nurtured to great effect. France, Britain's most powerful rival at the end of the eighteenth century, had no such tradition and even if it had possessed one before the Revolution it would have been obliterated quickly by the political reformers of the time.

Before the advent of the aeroplane and the missile, navies were the real measure of a nation's power. After the Napoleonic Wars, the two-power standard was initiated. This meant that to insure against attack from united enemies the British would always maintain a fleet at least equal to those of the next two most powerful nations combined. This naval superiority was sustained until the Washington Conference, a meeting of nations held immediately after the First World War, where it was agreed that the American Navy and the Royal Navy would be of equal size. But military dominance epitomised by naval power also meant that the most powerful nation on earth was also the most commercially successful, and that it was best placed to influence cultural and ideological change across the world. Sir Walter Raleigh understood this as early as the sixteenth century when he declared, soon after the defeat of the Spanish Armada: 'Whoever commands the sea commands trade, whoever commands the trade commands the riches of the world, and consequently the world itself.'

Yet to maintain its position the British military establishment had continually to modernise and strategists and planners needed to listen to innovators, inventors and designers. The success of the Industrial Revolution and the relative stability of the British political and social system provided a fertile environment for a new breed of engineers, scientists and proto-technologists during the eighteenth and nineteenth centuries. This meant there was no shortage of creative minds, but although the administration of the Royal Navy was an efficient and well-oiled machine, the British military was often held back by retrogressive commanders who preferred to stick with the old tried and tested ways. Consequently, during the middle of the nineteenth century the American military establishment, which was far more open to fresh thinking and innovation, began to take the lead. During and immediately after the Civil War, many military innovations entered the mainstream and rival nations, including Britain, were forced either to quickly adopt these advances or fall behind.

One of the best examples of the new breed of innovator was a man named John Ericsson. Swedish by birth, in his early twenties he emigrated to England where he worked on a variety of engineering projects, including designs for ship propellers and large-calibre guns and his own design for a steam boiler. During the course of this work he became friends with a senior US Army officer named Robert Stockton who was interested in applying the latest scientific and engineering ideas to the improvement of military machines. By 1840, Stockton had succeeded in persuading Ericsson to move to the United States, and there, through Stockton's contacts, he began to produce a succession of innovations for the US Navy and Army.

Ericsson's first revolutionary idea while in America was what he called his Caloric Ship, which he built with his own money and which he hoped would revolutionise marine technology. The unique aspect of this ship was the design of its boiler. Whereas standard boilers heated water to produce steam, Ericsson's machine worked on the principle of using the expansion of hot air rather than steam to move the pistons. In theory, the engine was more efficient and could move a ship much faster. Sadly, though, in this case, practice did not match theory. Ericsson's caloric engines suffered problems from overheating, and they were far heavier than the conventional equivalents, which meant that the improved efficiency of the engine was more than negated by their extra weight. In spite of great initial excitement over the caloric engine (including an offer from one keen entrepreneur of a million dollars for the patents), Ericsson's idea came to nothing and it was superseded a few decades later by the steam turbine patented by the English aristocrat Charles Parsons.

Ericsson had a mercurial mind and he was interested in every aspect of science and engineering. He also understood the power of bringing together different disciplines to solve a problem. Yet his first love was ships and he was especially determined to improve the capabilities and power of naval vessels. During the American Civil War he built a revolutionary battleship, the *Monitor*, for the Union forces, which had a revolving gun turret. The *Monitor* was later employed to great effect against a much larger and more powerful Confederate ship, the *Virginia*, in March 1862.

Ericsson was also instrumental in popularising the screw propeller. While in England he had learned a great deal about propellers from the experiences of an engineer named Francis Petit Smith who was working in London during the 1830s. Smith had been involved in an incident on the Paddington Canal in which his boat had collided with another, damaging its propellers. As he manoeuvred his boat away from the site of the collision, he noticed that it was travelling faster than usual. On returning to his workshop he discovered that the blades of his propeller had been sheared, so that they were now smaller. He therefore designed and built a new propeller, comprised of a large number of small blades, which was capable of moving his boat through the water at almost twice the speed of the original. Indeed, this design so impressed Isambard Kingdom Brunel that he changed the propellers of his ship, the largest of its day, the *Great Britain*, launched in 1843.

Although both Smith and Ericsson held different patents for screw propellers, it was Ericsson's work which made the most impression on navies all over the world. Ericsson's ship, the *Monitor*, was powered by conventional boilers but used his screw propellers, and after its successful engagement with the *Virginia* Ericsson went on to build a series of ships along similar lines for the US Navy before going on to experiment with submarines, torpedoes and heavy guns for use on battle cruisers.

In the middle of the nineteenth century two of the most important improvements in the propulsion systems and construction of marine vessels – the introduction of steam-powered ships and the use of iron-clad hulls – began to make a significant impact upon the navies of the world. Later, these developments would alter just as radically the designs and capabilities of merchant ships and other vessels. But, as is so often the case, the creative people behind these advances had to work very hard in order to impose their vision for the future of the navy upon policy makers and the men holding the purse strings.

What inevitably drove this change was a succession of arms races between Britain and her rivals. The military build-up of the French, followed later by the Americans, the Japanese and the Germans, forced British strategists and planners to accept innovation. Powerful and influential figures within the military and the government were very aware of the fundamental rule that if the enemy possesses any form of new revolutionary device or a military technique, then they too must have something at least as good, if not better.

Beginning in the 1870s, the navies of the leading powers were the centre of a succession of arms races which were forerunners of the rivalry between the United States and the Soviet Union during the second half of the twentieth century; for the time, these rivalries were every bit as expensive, beneficial to the advance of technology and crucial to maintaining the balance of world power.

Covetous of British naval supremacy, the French were the first to wade in. Their insecurity was amplified when, in 1840, the Royal Navy forced the French to withdraw from interfering in a conflict in Egypt. Powerful figures in the French government and armed forces felt that everything possible should be done to counter the Royal Navy. As a result, the French became increasingly receptive to new ideas. Soon after clashing with Britain in Egypt, some of the more adventurous strategists and military planners began to see the potential of using steam engines to power naval vessels. This seemed a perfect opportunity to take the lead against the old enemy; if French ships could outrun British vessels, they would naturally hold a huge advantage, or so it was thought.

The first vessel of this type had actually been an American creation. Robert Fulton, a determined and far-sighted engineer and inventor, had launched his first steam-driven vessel in 1807 and cruised the Hudson River in it. From this rather amateurish beginning the steam ship developed rapidly, so that little more than three decades after Fulton's maiden voyage steam ships were capable of crossing the Atlantic.

However, it was a commercial enterprise adopted by the military that pushed things forward with exceptional speed. In 1839 the British created a mail service to the Americas, and this innovation sparked huge competition between companies vying for exclusive carrier rights and for the government subsidies they would be entitled to.

This race pushed forward the development of faster and bigger ships. The leader in the field was the man who was perhaps the greatest engineer of the Victorian era, Isambard Kingdom Brunel. His ship the *Great Western* was the second to cross the Atlantic under steam. It arrived in New York twelve hours after a rival vessel, a paddle-wheeler called the *Sirius*, but it had made the journey far faster, having left port a full eighty hours after the *Sirius*.

The British government was quite aware of French interest in steam ships and with a keen eye to the military uses of such vessels they subsidised the ships built by Brunel and his rivals. One of the key stipulations in providing funding for these civilian vessels was that their design would allow for quick and easy conversion to battleships. In this way their thinking was reminiscent of the early days of naval development when all trading vessels could double up as battleships in times of need.

Although this system worked superbly and kept Britain just ahead of France in the Victorian age and beyond, many of the old school in the Royal Navy had to face a stark reality: either embrace innovation or allow their navy to be left behind. As late as 1828, just a decade before Brunel crossed the Atlantic in a ship partly funded by military planners, the Admiralty issued a statement in which they pronounced: 'Their Lordships feel it is their bounden duty to discourage to the utmost of their ability the employment of steam vessels, as they consider the introduction of steam is calculated to strike a fatal blow at the naval supremacy of the Empire.'[1]

It is easy to see why they should think this way. The very concept of innovation was alien to many powerful figures in the Admiralty. But others knew that this attitude would spell doom, that the worst way to deal with any threat was to bury one's head in the sand. The French, and later the Germans and the Americans, did not share the anxieties of some within the upper echelons of the Royal Navy's hierarchy. Indeed, they correctly viewed innovation, and in particular the power of the steam vessel, as their big chance. Fortunately for Victoria and her navy, and despite the Admiralty's stubbornness, enough entrepreneurs, influential politicians and strategists understood the danger and were not afraid to initiate change.

Yet even with the backing of these people, on two separate occasions the British were outstripped by the French, and for a short time the Royal Navy was posed a very tangible threat. With the launch of the *Napoléon* in 1850, the French demonstrated that they possessed the fastest warship on the seas. Capable of thirteen

knots, it gave the British such a fright that funds were found immediately to build a better, faster vessel in Portsmouth. A decade
later, in 1860, the French upped the ante again when they
launched a ship that could not only outrun any Royal Navy vessel
but also had a much better defence against enemy guns. The *Glorie*
was the first iron-clad battleship. The four and a half inches of iron
plate protecting its entire hull made it a revolutionary instrument
of war and it encouraged British naval engineers to work faster and
to compete with renewed vigour.

This 'Cold War' arms race continued for almost another half-
century as both nations spent vast fortunes on new designs, better,
faster engines and more heavily armoured battleships. Then, in
April 1904, the political and military map of Europe was transformed by an agreement forged between the British and French
governments called the Entente Cordiale. After more than a thousand years of rivalry and bloodshed, the two nations officially
became friends, for, apart from the obvious mutual benefits it
offered and the need to settle centuries-old colonial disputes, there
loomed the growing power of Germany which presented itself as
a tangible threat to the stability of Europe. Germany was rapidly
becoming the new pariah, a nation that was building up its military power with alarming speed and obvious intent.

The response to the German threat was inevitable: Britain and
France countered by beefing up their own military strength. The
most obvious result of this arms build-up was the creation of the
dreadnought class of battleship, a vessel that dwarfed all predecessors. Bristling with guns and torpedo launchers, clad in iron
and propelled through the water at a top speed of twenty-eight
knots using massive steam turbines, the largest of the class carried
eight thirty-eight-centimetre cannon each weighing one hundred
tonnes and capable of firing 885-kilogram shells a distance of
twenty kilometres. The dreadnoughts of the Royal Navy were like
nothing that had ever gone before them.

These vessels were incredibly expensive to build and maintain.
Only a small number were ever launched and they placed a serious burden on even the enormous resources of a British Empire

then pretty much at its peak. In his typically pithy style, Winston Churchill famously said of the trouble the navy faced in raising the funds to build these ships, '. . . in the end a curious and characteristic solution was reached. The Admiralty had demanded six ships: the economists offered four: and we finally compromised on eight.'[2]

As well as placing a huge financial burden on Britain, the introduction of the dreadnought also did much to accelerate the slide into war because it provoked the Germans into concentrating an inordinate proportion of their own resources into building up their fleet. Until 1906, when HMS *Dreadnought* took to the seas, the Royal Navy reigned supreme even in the face of German military escalation. By creating a small sub-fleet of seemingly invincible ships, they had changed the rules. These mammoth vessels immediately made the rest of the Royal Navy obsolete: to naval commanders and politicians alike, the only things that mattered were the dreadnoughts. The Germans quickly reached the conclusion that if they could build more dreadnoughts than the British it would give them an immediate advantage. The Kaiser was reported to have said of this new trend initiated by the Admiralty: 'You English, are mad, mad, mad as March hares.'[3] The first arms race of the twentieth century had begun.

In 1908 the Tirpitz Plan (named after the Secretary of State for the Imperial navy) to upgrade the German Navy appropriated funds to construct twelve dreadnoughts by 1912. This objective was met and surpassed, so that by the start of the Great War in 1914 the German fleet had thirteen dreadnoughts in full service and the Royal Navy had deployed twenty.

Historians argue about the significance of the dreadnoughts and their importance during the First World War. These ships were so expensive, were manned by such large crews and carried so much prestige and status that they were considered almost too valuable to be sent into battle. The loss of a single, precious dreadnought and crew would have struck such a terrible blow to the British Empire (or indeed to Germany) that the government was loath to use them. The one great exception to this was the Battle of Jutland in May 1916, when the British and German fleets clashed in the largest battle of the war and history's last great naval battle that involved neither aircraft nor submarines. The Royal Navy deployed thirty-three dreadnoughts and the German Navy eighteen.

The outcome of this battle was inconclusive. What is beyond debate, however, is that the entire venture of building dreadnoughts taught ship designers and engineers an enormous amount that was quickly used to make better civilian ships.

The largest of the dreadnoughts, HMS *Tiger*, launched in 1911, weighed 28,500 tons, and to propel such a mammoth vessel through the water at twenty-eight knots (several knots faster than any battleship a fraction of its size built up to this time) required extremely powerful and efficient engines. The building of a ship with such power pushed designers into developing the first oil-fuelled ships' engines. Without the dreadnoughts, the development of massive ocean liners, epitomised by the luxurious *Titanic*, the glamorous cruise ships of the years between the wars, and, most recently, stunning vessels such as Cunard's *Queen Mary II*, which made its maiden voyage in 2004, would have been much slower.

In 1922 and again in 1936, in a move prescient of the nuclear arms reduction treaties of the 1980s, the United States and Britain agreed to place an international limit on the size of battleships (but not on aircraft carriers). This agreement restricted any vessel to a maximum displacement of 35,000 tonnes, and at the same time gun size was limited to a forty-centimetre bore.*

But even as these restrictions were agreed upon and naval engineers reacted by designing sleeker, faster ships capable of carrying many different types of weapons, other innovations were about to reshape the navy and at the same time offer enormous potential for the explorer, the businessman and the scientist.

---

*Both these restrictions still made the maximum size and power allowed considerably greater than that of any dreadnought built.

Beneath the Waves

The most important of all the marine innovations in its effect on military and civilian life was the submarine. According to Aristotle in the fourth century BC soldiers of Alexander the Great used a submersible device to sabotage enemy ships. Two millennia later, Leonardo da Vinci drew designs for submarines, although it is only in recent times that models based on these plans have been constructed. In his notebooks he commented that he would deliberately suppress this invention '. . . on account of the evil nature of men, who would practise assassination at the bottom of the sea'.[1]

The idea of travelling beneath the waves has been a source of constant fascination, a dream as powerful as that of travelling to other worlds. Such excitement was captured beautifully by Jules Verne in his classic novel *Twenty Thousand Leagues Under the Sea*, written at a time when submarine development was just beginning to become practical.

Pioneering submariners had been experimenting with their machines for at least two and a half centuries by the time Verne wrote his novel in 1870. A Dutchman, Cornelius Drebbel, the court experimenter for King James I, claimed he had travelled beneath the Thames in a submarine, but there are no official accounts of this trip. A century and a half later, in 1776, a brave American rebel tried to blow up the British warship HMS *Eagle* in New York City harbour using a submersible called the *Turtle*, but his mission failed when he discovered that the ship had a copper hull to which he was unable to attach his explosive device.

The first marine enthusiast to produce a mechanically driven submarine was a Spanish maverick inventor and part-time lawyer named Narcis Monturiol. Although certainly a creative, brilliant and determined individual, Monturiol's important role in the development of the submarine has been largely ignored by historians for two very different reasons. First, Monturiol was anti-establishment, a political extremist whose socialist sympathies caused him to be exiled from his native Spain for a period. Consequently, he found it very difficult to acquire the finance and support he needed to build his submarines, and later it prevented him from being taken seriously by his government, even though his machines worked extremely well.

The second and most significant reason why Monturiol has been ignored is that he had no desire to see his invention used for military purposes. He was almost certainly the inspiration for the character of Captain Nemo, the misanthropic and anti-establishment commander of the *Nautilus* in *Twenty Thousand Leagues Under the Sea*, written a decade after Monturiol first conducted his experiments. An idealist, a pacifist and something of a well-meaning dreamer, Monturiol had started designing submarines in 1857 after he witnessed the drowning of a man employed to harvest coral close to the town of Cadaqués in southern Spain. His intention was to make the job of harvesting less hazardous by using a powered submersible, and he believed that such a device could also be of great service to the explorer and the scientist.

In 1859, Monturiol succeeded in launching his first vessel, the *Ictineo I*, which was built with private funds. Five years later he had designed and built a far superior submarine, *Ictineo II*, funded with 300,000 pesetas raised by enthusiasts across Spain and from Cuba who shared his vision.

Monturiol never received a peseta from the Spanish government or from any other official body, yet his submarine was a truly remarkable piece of engineering, which, if it had been taken seriously, could have advanced submarine technology by decades. He powered *Ictineo II* with an anaerobic engine (one that did not need air) that used a mixture of magnesium peroxide, zinc and potassium chlorate to generate heat to produce steam. A by-product of this chemical reaction was oxygen which was collected and used by the crew when submerged and which also powered lamps on the outside of the submarine.

Monturiol's ideas were remarkably advanced but they were not put to use again until almost seventy years after Monturiol's death, when the first naval vessel to employ an anaerobic propulsion system, the US Navy's first nuclear-powered submarine, the Nautilus SSN 571, was launched in 1954.

During the American Civil War submarines were used independently by both the Union and the Confederate navies, and it was through the widespread international reporting of this conflict that interest in the potential of the submarine began to grow. This war also saw the earliest use of the torpedo which quickly inspired naval commanders around the world. The combination of a submerged and therefore invisible vessel that could fire a missile underwater at a surface target was seen quite justifiably as a potentially devastating combination, a pairing of weapons that could alter radically the balance of naval power. When first shown what was described as an 'electric torpedo' (a form of primitive submarine), the then First Sea Lord, Admiral Sir John Jervis, declared presciently: 'Don't look at it. If we take it up, other nations will, and that will be the strongest blow against our supremacy on the sea that can be imagined.'[2]

The submarine has indeed fulfilled its potential and it has played a major wartime role since its first serious introduction during the middle of the nineteenth century. An example of its influence is the unfettered use of submarines by the German Navy to attack merchant shipping during the First World War. By 1915, the Royal Navy had achieved such success in creating blockades that the German population faced starvation. Germany reacted by attempting to destroy as many enemy civilian vessels as possible. This led to the sinking of the British liner *Lusitania* in May that year with such terrible loss of civilian life that it precipitated a change in the mood of the American public towards participating in the conflict. Within eighteen months, the United States had joined Britain and France in the war against Germany.

The submarine played an even more important role in the Second World War. Submarines accounted for only 2 per cent of the US Navy but they were responsible for sinking more than 50 per cent of the Japanese naval and merchant fleets. Both sides used the submarine to attack merchant shipping as well as naval vessels, and to protect their interests the British and American governments initiated a convoy system in which Allied shipping was escorted across the Atlantic by naval vessels.

Today, the submarines deployed by the major navies of the world are astonishingly sophisticated ocean-going vessels. Because oxygen for the crew is produced by self-contained chemical means and the submarine's engines are nuclear powered, they can stay submerged almost indefinitely and they can travel almost anywhere in the world, even under the polar ice caps. They also carry enormous firepower: since the 1960s, submarines have become mobile nuclear weapons platforms with the potential to utterly decimate an enemy nation without ever breaking the surface of the ocean.

The design teams which have developed the submarine have always been powerhouses of innovation. The reason for this is clear: parallel with the development of space vehicles, the ingenuity and technical skills required to build a machine that can safely transport a large crew, along with practical weapons systems, far beneath the surface of the ocean, is no ordinary achievement. The oceans are as alien as deep space and the only way humans can survive in this unforgiving environment is by extending their inventiveness and creativity to the limit.

All that has been learned from the development of these highly sophisticated weapons has also fed back into the scientific world. Today, the submarine is an invaluable tool for the explorer and the scientist. Robot submarines are employed to investigate deep-sea wrecks such as the *Titanic* that lie far beyond the range of a human diver. Deep-sea exploration has offered up a wealth of knowledge about marine life, and it has also taught us much about the geology of the ocean floor and about plate tectonics; consequently, our understanding of volcanology and seismology has progressed enormously in recent years, helping scientists to comprehend better the causes of earthquakes and how to predict them. The technology developed by naval designers in improving the submarine has also been of great benefit to engineers laying gas pipes and electrical cables. Submarines and robot submersibles can be used to inspect and repair cables and pipes that would otherwise be unreachable.

Another good example of military innovation that resulted in progress in civilian science is sonar. Invented as a military surveillance system, sonar has become almost indispensable to the marine biologist. Since the 1960s, many groups of sonar researchers working in the depths of the oceans have been funded by the US Navy and the Royal Navy. The military forces of both nations recognise that being able to identify sonar images produced by many thousands of fish species allows naval sonar operators to rule out natural echoes when they are searching for enemy submarines. Naval enthusiasm for such research was summed up by the US Navy's Rear Admiral John K. Leydon who commented: 'The Navy must continue to play a dominant role in the support of basic oceanography, in order that (1) major parts of the national effort go into those phases of the science which have the greatest Naval application; and (2) no broad area is neglected because of changing fads in the research community.'[3]

If they could somehow have seen the future, some of the men who first dreamed up the idea of a submarine and feared its misuse might have felt less apprehensive. The early undersea craft designers – men such as Leonardo da Vinci, the Englishman William Borne, who devised the ballast tank during the 1570s, and especially the Spanish inventor Narcis Monturiol – all began their experiments and drew their designs in order to create a machine for the scientist and the engineer, the explorer and the adventurer. Each of them made it clear he did not want to see his invention used for military purposes. Fortunately, although the submarine has become a powerful weapon and an invaluable tool to naval planners and strategists, it has also found many other uses in the hands of those who wish simply to learn rather than to kill.

PART 7

FROM THE

TRIBAL DRUM

TO THE

INTERNET

The story of how humans have progressed from the most limited form of communication to the ability to swamp the world with thoughts, images and words is a complex one and it follows a course that takes two distinct paths, strands that only truly intertwined late in the twentieth century. These came from twin drives – the need to know and the need to talk.

The part that military imperatives have played in trying to satisfy these two ambitions and to amalgamate the processes is difficult to overestimate. The need for communications on the battlefield or in organising a campaign is obvious, and the work of innovators who have created new and better ways to communicate has, with a few exceptions, been embraced by military financiers and governments throughout history. The need to know is never more pronounced than in the military arena where knowledge is power and power is the key to victory and survival.

The way in which war has influenced the development of human communication follows the classic pattern described throughout this book. That is, many of the innovations that mark the line of progress in this field came from commercial arenas where they languished in obscurity or suffered from lack of popular interest before coming into their own during a war, or in the lead-up to war. These innovations then returned, greatly enhanced, to the public stage after enthusiastic funding and commitment from military planners. Thus, we have mobile phones, microwave ovens, commercial aircraft guidance systems, radio and television, the microchip, the personal computer and nano-robots.

Smoke and Mirrors

For centuries the means by which man could communicate over a distance were extremely limited. The expression 'out of sight, out of mind' was quite literally true.

At first the problem was rooted in the fact that, with very few exceptions, the speed of communication was limited by the speed at which the fastest humans could move. Using a messenger, be it a runner or a horseman, was one of the few ways in which information could be passed from one location to another; and if those locations were days or weeks apart, then that was how long it took to communicate across the divide.

The ancient Chinese were renowned for their horsemanship and it is therefore perhaps not surprising that they relied upon relays of horsemen to transmit messages across the vast plains and through the mountain ranges of their land. Accounts tell of how riders could transport messages at a rate of three hundred kilometres (220 miles) a day.

The ancient Egyptians and the Babylonians used runners but the most famous account of one such messenger comes once again from Herodotus who described in *The History* the extraordinary feats of an Athenian named Phidippides who, as the legend goes, in 490 BC, covered a distance of 240 kilometres (180 miles) in a single night to enlist the help of the Spartans at the Battle of Marathon against the Persians. 'First of all, when the generals were still within the city, they sent to Sparta a herald, one Phidippides, an Athenian, who was a day-long runner and a professional,' Herodotus wrote. 'This Phidippides, being sent by the generals, arrived in the city of Sparta the day after he had left Athens. He came before the rulers and said: "Men of Lacedaemon, the Athenians beg you to help them; do not suffer a most ancient city in Greece to meet with slavery at the hands of the barbarians."'[1]

Whatever the truth of this episode, the use of runners or even horseman was clearly not the most efficient way of passing on news and from classical times many other forms of communication developed. By thinking laterally, the limit of human or animal speeds could be overcome, and although these methods were themselves far from perfect they did offer more versatility and scope for conveying important information.

One of the most inventive methods of communication used pigeons to carry written messages. Although this system was often unreliable as birds could easily fall into enemy hands, it continued to be employed by armies even up to the First World War.

But probably the most basic form of communication over a distance came from the use of sound and light. Primitive peoples lit beacons to signal important events, to muster armies or to report the death of a leader. Drums and horns were also used to transmit news from one group of people to another. But there were three obvious disadvantages to these methods. First, the message could only consist of a simple piece of information. Second, on their own, a single fire or audible signal could not carry news very far; and third, the conveyed message could never be kept secret.

The first disadvantage always placed limits on this method of communication, but another piece of lateral thinking helped overcome the second problem. The light from fires or the blasts of a horn could be relayed over any desired distance. One outpost would pass the message on to the next in a chain that could, in theory, span a continent. The problem of secrecy was a thornier

one, but the natural desire for privacy led to the invention of codes
and ciphers (see Part 3).

An important step forward came with the idea of signalling with
flags, mirrors and symbols which possessed the added advantage
that they could be encoded. The Greeks may have been the first to
employ these techniques. In his *Hellenica* the historian Xenophon
left a clear account of how the Greek navy used this method of
communication:

> On the fifth day as the Athenian ships sailed up, Lysander
> gave special instructions to the ships that were to follow
> them. As soon as they saw that the Athenians had disem-
> barked and had scattered in various directions over the
> Chersonese they were to sail back and *to signal with a shield*
> when they were halfway across the straits. These orders were
> carried out and as soon as he got the signal, Lysander ordered
> the fleet to sail at full speed. Thorax went with the fleet.
> When Conon saw that the enemy were attacking, he signalled
> to the Athenians to hurry back as fast as they could come to
> their ships. But they were scattered in all directions.[2]

This was obviously a very simple signal using a shield as a
mirror to reflect the light of the sun to a distant point, but this
primitive form of heliography was improved upon over the centuries, so that by using different lengths of signal and bunching
the flashes in certain ways it became an important means of communicating relatively complex information. Indeed, as late as the
1850s, the British Army was using an elaborate form of heliography to convey coded messages across the subcontinent of India.

Flags and semaphore could be used to transmit more information, and from ancient times up to the nineteenth century this
means of communication was often used in tandem with heliography. In around AD 80, the Roman historian Plutarch documented
the use of semaphore in his *Lives of the Noble Grecians and
Romans*, and almost a millennium later Byzantine naval commanders reportedly used flags and written symbols to send

information over large distances. According to the historian Francis Dvornik:

> During naval operations, the captains of the ships were expected to observe the *pamphylus* of the admiral, who gave orders by signalling from different sides and heights of the central flagship with banners of various colours, or with fire and smoke. A whole code of signals existed with which the commanders and their crews had to be acquainted. Part Nineteen of the strategic treatise ascribed to the Emperor Leo the Wise (866–912) gives numerous instructions as to the kinds of signals to be used and how the signalling should be handled. Unfortunately, the need for secrecy prevented the author from explaining the various signals then in use.[3]

Techniques such as those described here were of much greater use to sailors who could not rely upon land transport of any sort. This led governments throughout history to invest money and time in improving semaphore and other signalling methods and developing codes that commanders could employ to pass on, in secure form, essential and often secret information. Indeed, by the fourteenth century this system had become so ingrained in naval planning and training that a group of prominent English sailors produced a *Black Book of the Admiralty* which contained a list of signals and their meanings. Mechanical telegraphy was to mature very slowly over the centuries and a great deal of effort was expended by many engineers and scientists, all of whom believed their method of communication was better than that of their rivals.

One of the most important scientists of the late seventeenth century, Robert Hooke, a close contemporary and a great rival of Isaac Newton, was something of an intellectual magpie. Brilliant and energetic, he often took embryonic ideas from others and developed them much further. He was a man keenly aware of the *Zeitgeist*, and he knew how to extrapolate on a simple theme.

Hooke became fascinated with telegraphy, an enthusiasm matched by rival European scientists during the 1680s. On 21 May

1684, after working on a method of telegraphic communication for several months, Hooke delivered a lecture before the Royal Society where he was Secretary entitled: 'On Showing A Way How To Communicate One's Mind at Great Distances'. It began:

> That which I now propound, is what I have some years since discoursed of; but being then laid by, the great siege of Vienna, the last year, by the Turks, did again revive in my memory; and that was a method of discoursing at a distance, not by sound, but by sight. I say, therefore, 'tis possible to convey intelligence from any one high and eminent place, to any other that lies in sight of it, though 30 or 40 miles distant in as short a time almost as a man can write what he would have sent, and as suddenly to receive an answer as he that receives it hath a mind to return it, or can write it down in paper. Nay, by the help of three, four, or more of such eminent places, visible to each other, lying next it in a straight line, 'tis possible to convey intelligence, almost in a moment, to twice, thrice, or more times that distance, with as great certainty as by writing.[4]

Fine words, but Hooke's method, using a series of signs and coded letters and numbers all observed through the newly invented telescope (which he called 'tubular spectacles') was, in principle, little different from the methods of communication employed for centuries. And, although Hooke drew up elaborate plans for a chain of signal stations strung out across Europe and no doubt dreamt of the enormous rewards he would gain from such an invention, his methods were never put into practice and his ideas languished.

The fact is that all these methods of communication were unreliable and overly complicated. Bad weather could make signals indecipherable. Even coded messages could be intercepted and interpreted with relative ease; and the level of complexity of a message was still compromised even when using the most elaborate signalling system. Improvements came with the invention of the telescope, which allowed messages to be read over greater distances and during imperfect weather conditions; but by the middle of the nineteenth century the world was waiting for a completely different way of transmitting and receiving information. And this leap forward was facilitated by war.

Messages Through the Lines

When people talk about the telegraph they actually mean the electronic telegraph. From as early as the seventeenth century there had been innumerable attempts to create efficient and speedy communications systems using mechanical telegraphy and some of these had become popular.

Robert Hooke's design had been largely forgotten, as were most others, but in France a man named Claude Chappe dedicated his career to creating a pan-European network of telegraphy stations not dissimilar to Hooke's proposal and he became the first person to coin the phrase telegraph, or 'far writing'.

Chappe was born in the French town of Brûlon in 1763. He began a career in the Church before losing his benefices when the Revolution erupted thirty years later. He and his older brother Ignace, who was a government official, managed to persuade the new regime to finance a series of experimental signal stations, and in 1794 the brothers celebrated the first successful use of their device. By means of a mechanical telegraphic system they were able to communicate news from the battle front of the war between the French and the Austrians, informing Parisians of the capture of Condé-sur-l'Escaut from the Austrians less than an hour after it happened. The Chappe system was really just a more sophisticated version of the ancient methods of semaphore but using a clock face. The signaller set the hands of the clock to signify a number, letter or word and this was deciphered by a receiver using a telescope.

During the following decade, the fortunes of the Chappe brothers mirrored the rocky path of revolution, leaving them in and out of favour depending on who was running the country. At one point a confused mob, believing he was a spy, destroyed Chappe's equipment, but in 1799, when Napoleon Bonaparte seized power, he embraced the Chappe brothers' invention and financed a network of stations across France, establishing a telegraphic network linking Paris, Amsterdam and Milan.

This mechanical telegraph became fashionable for a while and Napoleon remained a supporter, but Claude Chappe was beset with self-doubt and became extremely sensitive to the criticism of his rivals. In 1805, he slid into depression and committed suicide by throwing himself down a well.

The first genuine advances from any of the adaptations of ancient communications ideas came during the 1840s and 1850s with the invention of the electric telegraph. This grew out of advances in the understanding of the nature of and interrelationship between electricity and magnetism.

Early in the nineteenth century, the Danish scientist Hans Christian Orsted discovered that a compass placed near an electric current showed a deflection of the needle. Although the theory behind this phenomenon was not fully understood until the end of the nineteenth century, enlightened men soon found practical ways to utilise the relationship between electricity and magnetism, and the invention of the electric telegraph became the first important development based upon this interaction.

The principle of the telegraph is a simple one. A tap or keying device, like a single typewriter key, closes and breaks an electrical circuit. The bursts of current and the length of the pauses between them represent letters and words. At the receiver's end, an electromagnet receives pulses of electricity according to what is sent, and this may then be translated back to the letters and words in the transmitted message.

In the earliest telegraphs the operator transmitting a message merely tapped out the pulses and this message was interpreted by a trained receiver. Later, the device became more sophisticated and included a sounder so that a receiver knew there was a message coming and an audio output could be used to sound out the message. Another variation saw a printer attached to the device and the on–off pulses of electricity wrote out a message.

The man credited with the invention of the telegraph is the American Samuel Morse. He had trained as an artist and was already a successful portrait painter when he made his name as an inventor. But instead of inventing the system from scratch, Morse actually took an old idea and improved upon it.

The concept of using electricity to transmit information had been discussed in scientific circles since the beginning of the nineteenth century, and at least one team of researchers in Germany had constructed an electric telegraph years before Samuel Morse. However, their system required twenty-six wires to connect the transmitter to the receiver, one for each letter of the alphabet. Morse took the fundamental concept (overheard during a conversation between two scientists on a transatlantic liner in 1832), simplified it and made it practical by devising a dot-dash code – Morse code – which became the accepted standard around the world.

As well as possessing an innovative mind and an eye for the practical, Morse was also a canny businessman. He was quick to exploit the military potential of his invention – that it could be used, for example, to transmit news between commanders in the field and headquarters – and as soon as he had a working model of the system he approached the US government for sponsorship. In 1842, Morse obtained a grant of $30,000 to 'wire' the country using his telegraph, and two years later, in May 1844, he became the first to transmit a message between two cities, Washington and Baltimore. The message, suggested to him by a family friend named Annie Ellworth, was a biblical quotation: 'What hath God wrought?'

Although with hindsight the attractions of electronic telegraphy might seem obvious, the public did not take to it immediately. During the first few years after Morse's experiments, telegraph companies and adventurous entrepreneurs struggled to make money from this new invention, but then, during the 1850s and 1860s, the Crimean War and the American Civil War transformed telegraphy into one of the most influential technologies of the nineteenth century.

These two conflicts were the first in which the electric telegraph was used to pass on information. Messages were encoded by generals at the front and sent to headquarters. Journalists filed news at the scene of a battle and wired it to their editors in the major cities. They were the first wars in which transcontinental telecommunications played a significant role.

In Europe after the Crimean War, and in the United States after 1865, there was an understandable sense of apathy towards military innovation, but the new communication device flourished in the commercial sector. By the end of 1865 there were 83,000 miles of telegraph wire in the United States alone, and Western Union, along with at least a dozen similar operators in Europe, had begun the commercial exploitation of the invention.

One of the earliest of these companies was the London District Telegraph Company which had actually been established in England two years before the beginning of the American Civil War. In its early incarnation in England, engineers had tried to put telegraph cables underground. In the United States, Morse too had initially followed this route and had reportedly spent two-thirds of his $30,000 grant on laying underground cables back in 1844. This method was ideal for keeping unsightly wires out of the way, and it kept the wires safe from the elements and from sabotage, but in cities like London and Paris it quickly proved to be prohibitively expensive.

This problem was solved in 1859 by an English engineer named Sidney Waterlow who had the original idea of running telegraph wires overhead, linking them by junction boxes on the roofs of houses and atop telegraph poles. This was a crucial step in substantially reducing the cost of creating an infrastructure for telegraphy and a development that made it commercially viable. Within a few years, news, share prices and even personal messages were being communicated through the system. At this time, the *Daily Telegraph* was first published, in its original form a purely telegraphic news service.

Soon after the American Civil War several countries established their own army telegraph corps. In Britain the Telegraph Battalion was established in 1884 and evolved into the Signals Corp and then the Royal Signals, which remains an important element of Britain's armed forces today. By the end of the nineteenth century, the world was crisscrossed by telegraph wires including transatlantic cables which, for the first time in history, enabled almost instant communication between America and Europe. Such innovations have led to the modern perception that the electric telegraph was 'the internet of its day'.

This assertion may be justified to a degree, but for all the improvements the telegraph brought it was still a pretty limited form of communication. Words could be sent and received initially by using code (and later via a printer or teleprinter which could offer recorded messaging) but what of the human voice? Was it possible, many engineers wondered, to devise a system through which people could actually talk to one another as if they were in the same room?

Well of course it was, and the man given the credit for pioneering the telephone was an American of Scottish descent named Alexander Graham Bell. Bell was born into a scholarly family and both his father and grandfather had been interested in sound. In his youth Bell built a succession of devices to receive and transmit vibrations. Apart from the family interest in acoustics, he was driven to this study by two additional factors. The first was an abiding interest in music (he was a gifted pianist); the second was the fact that his mother, Eliza, was almost completely deaf and Alexander was keen to find ways to help her detect the vibration patterns of speech.

Bell patented his telephone on 14 February 1876, just two hours ahead of a rival inventor named Elisha Gray who had developed a similar machine quite independently. During the next few years, Bell and his lawyers fought a series of court battles over the rights to the telephone. The case brought by Gray was the most significant of these and it went as far as the Supreme Court, where Bell won but settled out of court.

It is a striking fact that the military was barely involved in the early development of the telephone at all. Indeed, it is probably the most high-profile, universal device that did not originate from, or was in some way associated or linked with, military thinking. There are two reasons for this. First, military engineers of the 1870s simply refused to believe that such a device as the telephone was scientifically possible, and not until Bell made public demonstrations with his machine did they, along with the civilian observers of this new wonder, accept it. The second was that Bell had no interest in working with the military. From the beginning he saw his device as exclusively a business tool. Indeed, he was so wedded to this idea that he hated people even bothering to say 'hello' when speaking into his machine.* His thinking on the matter was so narrow he had absolutely no idea that the telephone would have almost universal appeal, and throughout his life he maintained an extremely elitist vision for the uses of his invention. 'The telephone,' he once declared, '. . . could not, and never would be an advantage which could be enjoyed by the large mass of the people.'[1]

When we consider this attitude coming from the very man credited with inventing the telephone it is perhaps not surprising that many others were extremely sceptical of it even after its successful public demonstrations. One British politician of the day declared in Parliament that the telephone was not needed because 'we have enough messengers here'. And during the telephone's

---

*Thomas Edison was similarly short-sighted when it came to practical applications for his inventions. When he produced his first phonograph in 1877, he published an article proposing ten uses for his machine. These included: recording books for the blind, preserving the words of dying people, teaching spelling and announcing the time. He did not mention recorded music at all.

very earliest days, Western Union claimed that the telephone could never replace the telegraph. An internal memo stated: 'This telephone has too many short-comings to be seriously considered as a means of communication.' However, the usually sceptical Lord Kelvin (who would later express the view that 'radio has no future') declared the telephone to be 'one of the most interesting inventions that has ever been made in the history of science', and that it was 'the most wonderful thing in America'.[2]

Some saw the telephone as nothing more than a gimmick, a novelty that would never go far beyond Vaudeville. Even the usually far-sighted Mark Twain failed to see its potential and turned down an offer from Bell to invest $5000 in the development of the telephone during the mid-1870s. Others could see how communications systems such as the telephone would soon change the world radically and were frightened by it. However, in the rush to adopt the device, and thanks to Bell's astonishingly rapid success, it was entirely overlooked that he was not in fact the real inventor of the telephone at all.

For well over a century history books have told the same story; that in spite of the many priority disputes that led him and his partners to the courts, Bell had been wholly original and the first to invent and to patent the telephone. This incorrect reading of history derived in part from the man's great showmanship. Within weeks of having his patent for the device approved, Alexander Graham Bell was holding huge public demonstrations of his machine in New York, where he connected his receiver to large loudspeakers so that his audience could hear messages sent from a transmitter placed several miles away. These displays had a huge impact and they made the inventor extremely famous very quickly. At the same time, such acclaim attracted financial backing, established what became the Bell Laboratories and the Bell Corporation and made the inventor of the telephone very wealthy. This success was driven partly by American politicians and business people who took to the telephone almost immediately. Europeans too were quick to see its potential. Also, thanks to the enormous popularity of the telegraph, the infrastructure for long-distance communication was already in place and so the telephone could be absorbed quite easily into the system.

Yet it was by sheer fluke that Bell followed the path to fame and fortune, and it has only been accepted in recent years that an almost entirely unknown Italian inventor named Antonio Meucci (who died in poverty) was actually the man who had documented, recorded and built telephones more than five years before Bell made his first working model.

Born in 1808, Meucci studied engineering and design at the Academy of Fine Arts in Florence, thereafter spending most of his adult life travelling the world applying his engineering genius wherever he could. But in his spare time he expended huge efforts inventing devices that could transmit and receive sounds rather than simple telegraphic messages.

In 1850, he settled in New York with the specific intention of playing a part in the new innovation and technology boom for which America was becoming recognised around the world. Unfortunately for Meucci he had absolutely no head for business and he and his wife became so impoverished that they were forced to sell his early model telephones for less than $6 to help feed themselves for a few weeks.

In 1871, Meucci took the advice of a friend and scraped together enough money to take out a patent on his telephone and he made an approach to the American military with an offer to demonstrate his device. The demonstration did take place but the officers sent to observe it filed a report in which they claimed Meucci's machine was nothing but a clever deceit.

This proved to be just one of many setbacks for poor Meucci. Soon after this failure, his financial situation became so dire that, after paying patent administration fees covering the period between 1871 and 1875, he found he could no longer afford to renew his patent for the telephone and at the beginning of 1876 he let it lapse.

At this time Meucci was working on perfecting a number of electrical devices and he rented a cheap laboratory with several other inventors and amateur scientists. He shared one of those labs with a then unknown twenty-eight-year-old with a keen interest in all things related to the transmission of vibrations. His name? Alexander Graham Bell.

Whether Bell was aware of Meucci's financial difficulties and the fact that he had allowed his patent to lapse has never been confirmed beyond reasonable doubt, but the events surrounding Meucci's loss of priority are rather mysterious and Bell lodged his patent just weeks after Meucci's lapsed. To add spice to the story, in 1874 the Italian inventor had sent a model telephone along with plans and designs to Western Union. He heard nothing from them and a few years later, when Meucci decided to take legal action against Bell, investigators went to the Western Union offices where they were told that all materials pertaining to Antonio Meucci and his designs had been lost.

Infuriated by what he saw as an obvious cover-up, Meucci decided to switch his claim against Bell from a question of priority to one of fraud. He was encouraged by the admittedly suspicious fact that only a few months before Western Union reported that they had never heard of Antonio Meucci's telephone they had signed a deal with Alexander Graham Bell that would make all concerned extremely rich.

Meucci had supporters, though, and he secured a verdict from the Secretary of State in which it was declared: 'There exists sufficient proof to give priority to Meucci in the invention of the telephone.' But even this clear support helped little. Thanks to red tape and bureaucratic delays, the fraud trial was constantly postponed, until finally, in 1888, a trial date was fixed for the following spring. Early in 1889, however, just a few weeks before the case against Bell was to be heard, Meucci died aged eighty and the charges of fraud against the globally acclaimed Bell died with him.

However, this dispute did not end with Meucci's death and he has been granted some form of posthumous victory. For more than a century, Italian campaigners and certain historians have struggled to have Meucci's name highlighted and his achievement recognised. It has proved to be an uphill struggle because the name of Alexander Graham Bell is so famous and the weight of history has, in this case, become quite immense. Yet, against the odds, the revisionists have triumphed. After intense lobbying, in June 2002 US Congressman Vito Fossella succeeded in having passed House of Representatives Resolution 269, 'Sense of the House: Honoring the Life and Achievements of 19th-Century Italian-American Inventor Antonio Meucci'. This resolution means that, officially, Alexander Graham Bell can no longer be credited with the invention of the telephone and that henceforth this honour should instead be given to a little-known Italian inventor called Antonio Meucci.

Sadly, though, this judgement does not alter the fact that the birth and early life of the telephone were controlled completely by Bell and his associates. Perhaps if the army officers who claimed to have studied Meucci's device had seen its potential things would have been very different, but Bell, who was determined to see the device used solely for business purposes, ensured that the military had very little influence over the way the telephone developed during its first decade. As a consequence, it was not until the Boer War, which began in 1899, that the armed forces started to take the telephone seriously.

One of the earliest pioneers of the field telephone (as the military version became known) was Lars Magnus Ericsson, who developed the device in the 1890s and went on to sell several models to the Swedish Army before landing a large contract with the British government. However, Ericsson's interests soon moved from the conventional telephone, using wires strung out over a distance, to 'wireless' or radio communication, and he became the earliest pioneer of what has become known as the mobile phone.

Ericsson was also the first to suggest fitting a telephone system into a car. This involved the driver stopping close to a telegraph pole and connecting wires from the car to the transmission lines. According to the official history of the Ericsson Company, the inventor and his wife, Hilda, enjoyed cross-country driving and Lars invented a car phone so that he could keep in touch with his business interests: 'In today's terminology the system was an early "telepoint" application: you could make telephone calls from the car,' the company's account tells us. 'Access was not by radio, of course, instead there were two long sticks, like fishing rods, handled by Hilda. She would hook them over a pair of telephone wires, seeking a pair that were free . . . When they were found, Lars Magnus would crank the dynamo handle of the telephone, which produced a signal to an operator in the nearest exchange.'[3]

This sounds fantastically quaint, but in 1901, when Ericsson came up with this idea, there was no other alternative and almost half a century was to pass before the first genuine car phone appeared. This was in 1946, when the Swedish police force installed a test model in a car. It could make no more than six calls before the battery expired, and it was not until phones could be linked via satellite and a network was created offering a far greater range and coverage that the mobile phone began to be seen as a practical device.

The modern mobile phone was developed for military purposes by Britain's DERA (the Defence Evaluation and Research Agency) during the 1960s, and it was not until 1979 that the first commercial network began in Japan. Its successors, now used globally, have reached a high level of sophistication. Today, some 75 per cent of Westerners have mobile phones (although Americans are the least interested in the device, with only 66 per cent of the population using what they call a 'cell phone'). One of the global leaders in mobile phone technology is the Swedish company Ericsson, founded by Lars Magnus Ericsson.

Elements of the armed forces were certainly slow to adopt telephones and the story of Meucci's failure is far from unique. But in some ways it is easy to see why the military became enthusiastic about utilising some form of telephone rather later than the general public. It must be remembered that a telephone can only operate by transmitting a signal along a wire. This meant that although the technology was available to the military years before the outbreak of the First World War, telephones were rarely used in battle because they were quite impractical. Until the invention of wireless phones and walkie-talkies, the primary military use of a conventional telephone was to allow communications between headquarters and stations some distance from the front. In addition, the early field telephones were heavy and unreliable and did little to inspire military men to abandon tried and tested methods of communication such as the telegraph, pigeons, fires, flags and runners.

Because of these limitations, the conventional telephone never did become essential to armed forces. Instead, the military establishment embraced another invention which first appeared a little over two decades after Alexander Graham Bell had become a household name.

Words Through the Ether

Fights over priority are far from unusual and the story of how the radio was invented and came to be used globally offers another example of just such a battle. However, there is no debate over who discovered the theoretical basis of radio. In 1865, the great British scientist James Clerk Maxwell predicted the existence of waves similar to light but of different frequencies. One region of the electromagnetic spectrum he described consists of radio waves which travel at the speed of light and may be reflected, refracted, transmitted and received just like visible radiation.

Less than two decades after Maxwell's discovery, the German Heinrich Hertz demonstrated this wave theory in laboratory conditions in which he produced radio waves and other radiation and studied their properties. These researches resulted in his 1892 landmark work, *Untersuchungen Ueber Die Ausbreitung Der Elektrischen Kraft* (*Investigations on the Propagation of Electrical Energy*). Hertz's book was read widely by engineers and those interested in new science and within a couple of years of its publication scores of inventors and researchers around the world were investigating the uses of electromagnetic radiation, and in particular the possibilities offered by radio waves.

With the exception of one essential component a radio works in the same way as a telephone, or, for that matter, an electrical telegraph. In all three systems vibrations are converted into a transmittable form and a receiver at a distant point translates this signal back into the original vibration. In a telegraph this signal consists of dots and dashes passed through a wire. The telephone advanced this system by allowing sound, and in particular speech, to be encoded, passed through a wire and decoded at the receiving end. Radio is the same except that the signal is transmitted through the air rather than along a wire. The radio transmitter translates sound into an electromagnetic impulse that lies within the radio frequency region of the spectrum. This impulse is transmitted through the air to a receiver where it is converted back into a sound which may then be amplified using a few basic electronic components.

Between the 1860s and 1890s, dozens of patents were filed for systems that purported to use radio waves to communicate at a distance. Inventors offering designs for radio devices included such luminaries as Nikola Tesla. Another enthusiast was an engineer named Mahlon Loomis who claimed to have conducted successful experiments in the transmission of radio waves during the American Civil War, a decade before Hertz conducted his laboratory experiments. Loomis patented his ideas but his invention was left to fade into obscurity for lack of financial backing.

Because of intense rivalry to produce an operable radio transmitter and receiver, and thanks to the deluge of patents and published ideas, it is almost impossible to say who was first to create a workable radio system. The name that stands out in the field is, of course, Guglielmo Marconi who began serious experiments into radio devices at his home in Bologna during 1895, when he was twenty-one. That year, Marconi delivered his findings to the Italian government and offered to perform a demonstration for them, but his ideas were received with nothing but apathy and scepticism. Undeterred, he decided to take his invention to England where he arranged a demonstration for the Admiralty.

That radio could be of particular use to the navy seemed perfectly obvious to Marconi. By the 1890s, the telephone had become extremely popular amongst those who could afford it, and although its potential on the battlefield was extremely limited it was in regular small-scale use with the armed forces of the world for land-based communication. However, at sea the telephone was completely useless. At the same time, the need for a fast, efficient and secure means of communication was on the wish list of every navy, limited as they were to mechanical telegraphy in its many forms.

Luckily for Marconi a few key people within the Admiralty in Britain were more open to innovation than the Italians. The Royal Navy was the most powerful in the world and it spearheaded the military colossus that sustained the Empire. As we have seen, older admirals and senior administrators often dragged their heels when it came to embracing new technology, but Marconi managed to find just the right people to help him gain attention for his radio, a device that was quickly dubbed 'Marconi's magic box'.

In February 1896 Marconi reached England where he almost immediately found backers willing to finance the further research and development of his radio apparatus. In fact he was so successful at attracting attention for his invention that within little more than a year of setting up in business in England he was fighting off potential investors, half a dozen at least, all wanting exclusive rights. These included the US Navy, a group of European private businesses and the British Post Office.

Military rivalry dictated who was the first to acquire radio as a practical device. An Englishman named William Preece, who was the Chief Engineer for the Post Office, had for some time been conducting his own experiments into a practical radio system and when he learned of Marconi's work he arranged a demonstration for his colleagues and superiors. At this demonstration in May 1897 Marconi succeeded in transmitting a radio signal over a distance of three miles, an event now viewed as the first documented, successful radio communication. Present that day was a German professor by the name of Adolphus Silby, who, fired up by what he had witnessed, returned immediately to Germany to inform his government of the amazing new device.

During the 1890s the German military build-up was beginning to present a very tangible threat to the British Empire, and it did not take the British government long to learn that Silby was already attempting to duplicate the experiments he had seen performed by Marconi, nor that the German government was funding him and others to create a rival radio system as quickly as possible. With no time to lose, in the summer of 1897 the Royal Navy offered Marconi an exclusive contract to build radios for the fleet.

The revolution in communications that radio initiated cannot be exaggerated. Although the Royal Navy tried to keep details of Marconi's work secret, it proved an almost impossible task. By the end of the nineteenth century, all the world's major navies had acquired the system for ship-to-ship and ship-to-shore communication.

Strangely, perhaps, Marconi and almost all the other influential figures working on the early development of radio had no idea of its vast potential as a medium for mass entertainment. The very notion of broadcasting anything other than military messages, distress calls from ships or government information appears never to have entered their minds. This lack of vision was principally due to the fact that for some time no one was actually using radio to transmit the spoken word; rather, all the early communications using the device employed Morse code. To the pioneers of this technology the only significant advance radio offered was that it was wire-less.

After the existence of radio became known to the public, a number of entrepreneurs quickly grasped its vast commercial potential. The first voice transmission was made in 1906, over a distance of eleven miles between a station at Brant Rock, Massachusetts, and a ship anchored in the Atlantic, and this feat attracted innovators ready to exploit radio on a commercial level. Suddenly, businessmen became very interested in the possibilities of broadcasting news, speeches, music and talks.

In fact, media people became so quickly taken with the potential of radio that governments were forced to introduce strict controls on transmissions because signals filling the air were interfering with each other, threatening to destroy the broadcasting industry before it had properly begun.

The dangers represented by unregulated broadcasting were illustrated horribly in April 1912 when the *Titanic* sank en route to New York, claiming an estimated 1500 lives. The ship was equipped with a radio but the signal and those of potential rescue ships, all clamouring to help, caused such interference that it actually hampered the rescue operation and compounded the tragedy.

Radio had been used with only very limited success during the Boer War of 1899–1901. In fact, it was considered so unreliable that journalists filing copy from South Africa refused to send their reports home using this method. But by the time of the First World War, only a decade and a half later, radio had become essential technology for military planners and strategists. During this conflict, the British and American governments, along with their German counterparts, each established a government monopoly on radio and they carefully controlled transmissions. This stifled civilian use of radio until 1918, but, with the end of hostilities, many of the technicians, engineers and signalmen who had worked with radio during the war returned to civilian life with fresh ideas for its commercialisation.

The first commercial boom in radio occurred between 1921 and 1925 when large numbers of radio sets were made and sold at prices that allowed the average family to own what became known as a crystal set, or wireless. Companies sprang up to make programmes to fill the airways and radio became a staple of everyday life. Very soon after the end of the First World War the British and American governments created a relatively liberal licensing scheme for broadcasters, but soon there was such interest in obtaining broadcasting rights that the regulations were revised because of fears that so many radio transmissions would interfere with the armed forces using their crystal sets. This led to the allocation of different radio frequency bands for specific purposes. The military operated in one part of the radio spectrum and civilian broadcasters were limited to another. By 1922, a consortium of companies, which included the rapidly growing Marconi

Company along with military suppliers such as Vickers, GEC and Western Electric, created the BBC, which first broadcast on 14 November of that year.

Radio also quickly cross-fertilised with the telephone so that, by the Second World War, wireless field telephones were in common use. The difference between these telephones and the earliest devices used with little success or enthusiasm during the Boer War and the First World War was that they were much smaller, had a practical battery life, and, most crucially, did not require any form of wire running between transmitter and receiver because they operated using radio frequencies. Of all the types of field phone used, the walkie-talkie was the most popular and it is still used extensively by the military and emergency services.

Today, the radio remains the mainstay of almost all communications systems. The mobile phone is a close relation of the walkie-talkie, or the two-way radio; more accurately it could be described as the offspring of the telephone and the radio. Meanwhile, although television has become a far more powerful force in the field of entertainment and broadcasting, radio still remains an immensely popular and important medium and it offers its own unique qualities that cannot be superseded by television.

From Death Rays to Microwaves

As television demonstrates, radio waves are not the only form of electromagnetic radiation that can be transmitted through the air and this is the basis of one of the most useful military innovations of the mid-twentieth century.

The development of radar is a prime example of how ideas that may languish through lack of support receive a remarkable boost from the military establishment during times of war. Often this success is belated because the military may have missed an opportunity years before when the pressure to act, to improvise and to take some risks was less intense. Long before the potential of radar was finally realised, its concept had passed through military hands at least five times in different countries only to be rejected each time.

The principle of radar (a naval acronym for 'Radio Detection And Range-finding') is remarkably straightforward and it derives from the ideas of James Clerk Maxwell and the practical discoveries of Heinrich Hertz. Hertz conducted experiments in which he transmitted electromagnetic radiation the length of his lab and bounced them off a carefully placed object. He knew the speed of light, and so, by recording how long the waves took to travel to and from the object, Hertz was able to calculate the distance between the transmitter and the object. This was really the very first use of radar. It was an extremely primitive version, but, because of this crucial experiment, if any individual should be given the credit for inventing radar it should be Hertz.

Heinrich Hertz was not slow to understand the potential of his discovery but others certainly were. He demonstrated his experiment to a group of naval strategists and proposed that his system could be used for the detection of ships, as a navigation aid and as a life-saving procedure for vessels in distress. He was ignored and nothing came of the demonstration. This same story was repeated a few years later when fellow German Christian Hulsmeyer suggested using radio echoes to help avoid collisions at sea and presented his case to a different group of naval experts.

The fate of the *Titanic* convinced scientists of the need for some way of detecting the presence of other ships, icebergs or land masses, especially during bad weather when visibility was greatly impaired. If the *Titanic* had possessed such a device, so the theory went, it would never have sunk, and some 1500 lives could have been saved.

Scientists understood this argument but the money men did not. Even in 1922, ten years after the *Titanic* tragedy, the US Navy still saw nothing of value in a demonstration of a primitive radar system by two respected military researchers, Albert Hoyt-Taylor and Leo Young, who worked at the Naval Research Laboratory in Washington. Eight years later, the two scientists tried once again to interest the military establishment in their experiments but they failed again. It was left to the British to make radar a success; and the threat from Hitler's Germany acted as a powerful catalyst.

At the start of the First World War the aeroplane was barely a decade old and its primary uses were as a reconnaissance weapon and for occasional aerial attacks launched from makeshift airstrips close to the front. However, by the mid-1930s, the potential of bombers for attacking enemy cities hundreds of miles away awoke politicians to the danger presented by enemy air forces. Questions were raised in Parliament and in the Reichstag. In the Commons in 1932 the Prime Minister made the prophetic statement that there could be no way to stop an attack from bomber aircraft and that the only logical response was to do the same to the enemy. Within two years, military manoeuvres had demonstrated that, should war break out, RAF bombers could reach targets in Germany.

A very real example of just how powerful the aeroplane had become can be seen in the photographs and film footage taken during 1936 when Nazi bombers caused terrible loss of life and widespread destruction in the Spanish Civil War. In particular the devastating bombing of Guernica sent a shudder through Great Britain's military establishment, and, although many officials remained oblivious to the dangers that faced the country, a few saw all too clearly that British air defences were woefully inadequate and left the country wide open to attack from the air.

A discussion and planning group called the Committee for the Scientific Survey of Air Defence (CSSAD) was established and headed by one of the most notable scientific figures of the day, Sir Henry Tizard. However, secret groups working within the defence establishment had for some time been considering using radio waves and other radiation to create some as yet unspecified defence or attack system.

The first idea of the CSSAD, which probably knew nothing of these secret groups, was to consider the possibility of building a wave system that could be used to bring down enemy bombers. This led to the idea of what quickly became known as a 'death ray'.

It sounds terribly quaint now, almost like something from an Ealing Comedy, but during the late 1930s the idea of a death ray was taken so seriously that a prize of £1000 was offered to any inventor who could demonstrate the power of such a ray by killing a sheep at a distance of two hundred yards. One man who had actually been pursuing the idea vigorously for at least a decade before this was an eccentric inventor-cum-charlatan named Harry Grindell Matthews. Matthews became something of a celebrity during the 1920s and early 1930s and he was fêted by the media of the day who dubbed him 'Death-Ray Matthews'.

The story of Harry Matthews is a tragicomic affair in which our hero spent most of his life on the fringes of the law, existing as a con man who always managed to stay one step ahead of those who could expose him as a fraud. Typical of his breed, he succeeded in tricking a few important people into believing his plans, including a scheme to construct a workable death ray. But, in spite of making (and losing) a small fortune from his efforts, he never came close to producing a practical weapon that could bring down an enemy aircraft using any form of electromagnetic radiation. His was really a sideshow.

The real action was taking place amongst orthodox scientists, whom the Ministry of Defence employed officially in 1935. As mentioned earlier, there were already clandestine groups working since at least the mid-1920s on the potential of utilising radio waves and other forms of electromagnetic radiation for military purposes. There is evidence to suggest that these researchers had considered the idea of some form of electromagnetic weapon but that they had given up with this particular line of investigation.

One of the most advanced research teams working in this area included the man who was later to be credited with the invention of television, John Logie Baird. Baird had conducted his first demonstration of a working television system in a lab in London's Soho in January 1926, but within two years, and as he was refining his device, he was also working for the Ministry of Defence on top secret plans to use radio signals and other electromagnetic radiation for non-specific military purposes.[1] Little is known about what came of these ideas except that when the MOD contacted another highly regarded scientist working in the same field, a Scottish physicist named Robert Watson-Watt, he knew better than even to contemplate a form of death ray. He immediately informed the ministry that such a thing was impossible and that in terms of defence the most useful thing to do with electromagnetic radiation was to construct some form of early warning system.

The principle behind Watson-Watt's radar was almost identical to that of others before him, men such as Hertz, Huelsmeyer, Hoyt-Taylor and Young. But Watson-Watt was the first scientist to make the device practical, and, because of this, he is considered something of a military saviour, the man who, by providing the RAF with a huge tactical advantage during the Battle of Britain, did more than most to save the country in 1940.

Watson-Watt's idea was simply to sweep an expanse of sky with a beam of electromagnetic radiation. If the beam hit an object it would bounce back and be picked up by a receiver. This was, in theory, a simple matter. Maxwell had described the theoretical foundations of the idea, Heinrich Hertz had shown it was possible under laboratory conditions, and, by the 1930s, radio was an everyday technology. However, turning theory and lab work into practice in the real world was quite another matter. Knowing where to sweep, knowing what was an enemy aircraft and what was some other distant object, and developing enough power in the beam to provide a workable range and accuracy were all fundamental problems that were extremely difficult to solve.

It took almost a year to produce a test model and even then it was thanks primarily to a handful of enthusiasts that sufficient funding was found to build an effective radar defence system. One of the key figures behind this move was Air Marshal Hugh Dowding, a rare visionary within the military establishment. Dowding was an individual, unorthodox, opinionated and determined, as a result of which he was greatly disliked by many of the top brass. Fortunately, though, his talents were appreciated by Winston Churchill who gave Dowding his personal support in establishing a research facility which appropriated almost unlimited funding and was given top priority during the years immediately before the start of the Second World War.

Thanks to this initiative, Britain became the only country with a workable radar system a year before the outbreak of war in September 1939. This defence system consisted of a chain of stations, each of which probed a section of British airspace in the south and east of the country, the regions most vulnerable to attack from mainland Europe. By 1940, when radar was first put to use in combat, during the Battle of Britain, other less advanced radar systems were under construction in Germany and in the United States and radar was already in regular use by the Royal Navy.

This early radar system was a very simplistic affair. Known as continuous-wave radar, it worked on the principle of transmitting a continuous electromagnetic signal and detecting the reflected waves returning from a solid object. It was limited by the fact that as waves travelled on past an aircraft they could make contact with more distant objects and confuse the radar operator. With these early devices it was also very difficult to distinguish a genuine signal from reflections produced by buildings or other presences on the ground. Added to this, the device was unreliable in bad weather.

A vast improvement came with the introduction of pulsed wave radar. This works on the principle of sending a discrete pulse of electromagnetic radiation in a narrow beam which, like the continuous wave, bounces back if it encounters an object. The difference between this system and continuous wave radar is that, because a burst of radiation is used, it does not become confused with random reflections and produces an uncluttered radar image of an enemy aircraft.

The creation of this more advanced system relied on a revolutionary piece of equipment called a resonant-cavity magnetron, first built in 1939 by two other British scientists, Henry Boot and John Randall. This machine enabled a new form of radar to be constructed which used higher frequency radio pulses called microwaves, which allowed for more accurate detection and a far greater range.

By 1940, Boot and Randall had improved the performance of their machine (already nicknamed the 'magic eye') so that it could produce a pulse one hundred times more powerful than that produced by the largest conventional continuous-wave transmitter. However, by this point in the war Britain was struggling to survive. The continuous-wave radar system had been put in place at enormous expense and resources could not be stretched to establish a new network, even if it offered a vast improvement upon the system already in place.

Undeterred, Boot and Randall, along with Sir Henry Tizard and a small team of other scientists, flew to Washington with their device and with American money they established a research and manufacture facility to begin constructing the new radar system. The work progressed rapidly and by the end of 1941 it was in operation in Hawaii where a pulse radar succeeded in detecting Japanese planes an hour out of Pearl Harbor, although it didn't do them much good.

Much has been written about the tremendous importance of radar to the Allied victory in the Second World War and enthusiasts of the technology are prone to hyperbole. One example is a book by Robert Buderi (otherwise extremely good) about the invention of radar called *The Invention That Changed the World: How a Small Group of Radar Pioneers Won the Second World War and Launched a Technological Revolution*. Having said that, there is no denying that radar served a very important role throughout the war. The Allies were years ahead of the Axis powers in developing this technology and that advantage was never lost. Radar was put into aircraft and ships and a variation of the technique was used to develop sonar used by submarines and by submarine hunters. Towards the end of the war, the British used newly installed pulse-wave radar to help detect V1 and V2 rockets on course for London and other populated centres in the south of England.

After the war, radar technology developed rapidly and assumed a pivotal role in ship and aircraft navigation systems that were only partially superseded towards the end of the twentieth century when satellite guidance systems and GPS became operational. Radar still plays a key role in early warning detection and in monitoring shipping lanes and commercial aircraft flight paths.

One other spin-off from radar is the humble microwave oven. In 1945, a researcher named Percy Spencer, working on improving the design of the resonant-cavity magnetron, accidentally stood in the beam close to the transmitter. He was unharmed but noticed that a bar of chocolate in his shirt pocket had begun to melt. His curiosity piqued, he sent out for some popcorn. Placed in the beam, the popcorn exploded, sending pieces across his lab. Next, he placed a raw egg in the beam. This too exploded as the inside cooked first and expanded rapidly. These experiments intrigued him and he began to think about the possible applications of this serendipitous discovery.

Spencer's employer was the Raytheon Company, one of a handful of businesses that had sprung up to reap the rewards offered by the invention of radar, and their primary income came from supplying the armed forces. After the war, the demand for radar plummeted for a while during the period before commercial applications became commonplace.

Perhaps because his first experience with microwaves had come from heating food, it struck Spencer almost immediately that the waves could be used in the kitchen. It was an idea that captured the imagination of executives at Raytheon and the company began to develop the first microwave oven, acquiring the US patent in 1953.

The Raytheon microwave was a huge, power-guzzling and very expensive machine (retailing for $3000 in 1953). However, it was intended not so much for use in the home but in restaurants, aboard ocean liners and on transcontinental trains. The job of creating a domestic version that was compact and cheap fell to the Japanese who marketed the first one in 1955. It was slow to catch on and it was not until the mid-1960s that the microwave was embraced by the public. Today, more than 80 per cent of Western homes have a microwave oven.

Numbers

The military have always been keen on numbers and, when we consider that many of the great battles in history have involved the movement and organisation of some of the largest numbers of human beings gathered together in one place, it is easy to see why.

Numbers are also essential to military planners who need to know statistical facts about weapons and other supplies, as well as the number of personnel in each location in the theatre of war. At the same time, without at least a basic understanding of mathematics, the direction and distance of a target is almost impossible to ascertain.

Applied mathematics comes in a multitude of guises and it helps the military in many diverse and often indirect ways. One of the best examples of this is the application of mathematics to the science of meteorology, which has become both an important civilian discipline and an invaluable element of military strategy.

Weather prediction was conceived in the eighteenth century and developed in tandem with ballooning. One of the earliest uses for balloons, along with surveillance of enemy forces, was the study of changing weather patterns. For centuries ships had been lost at sea and military campaigns ruined because the planners of attacks could never predict the caprices of Nature. With the advent of the balloon it was possible to make long-range observations which could then play their part in judging the best moment to launch an offensive.

This was a valuable advance, but in the early days of meteorology very little was known about the forces of Nature responsible for shaping weather. Today, meteorology is recognised as an extremely complex science and those who study and forecast the weather depend on computers to make their predictions. The weather changes for many different reasons; it is a classic example of chaos theory in action.

However, mathematics may be used to model weather patterns and the more information a mathematician has the better it may be used as a predictive tool. Lewis Richardson, a British ambulance driver with the Quaker-run 'Friends Ambulance Unit' during the First World War, knew this. During breaks between rounds of duty he sat with notebook and pencil and tried to elucidate as many equations as he could governing the numerous factors that make the weather what it is. He considered the way the sun heated the earth, he factored in wind speeds and the variation in pressure from one location to another and he studied temperature gradients.

This took him a year and at the end of it he had completed a detailed series of calculations which offered what he believed to be an accurate way of predicting the weather up to forty-eight hours in advance. Then, in April 1917, just as he was about to pass his work on to his commander, Richardson's manuscript was lost under a pile of coal during the chaos of the Battle of Champagne.

Miraculously, a year later, the manuscript was discovered and in the spring of 1918 Richardson offered his work to the British High Command. At first the senior officers dealing with Richardson's proposal were dismissive, declaring: 'The British Army fights in all weathers.' However, once this scepticism was overcome a small team was assigned to test the method. Unfortunately, Richardson's equations proved hopelessly wrong and the military meteorologists were so discouraged that they refused to consider the work any further and reverted to traditional methods. It was only immediately after the Second World War and the advent of the computer that scientists revisited the thirty-year-old work, and, using Richardson's equations with greater care and employing the enormously improved number-crunching power now available, they concluded that discarding the original work had been a mistake. Today, meteorological calculations carried out by some of the world's most powerful computers are based on Richardson's equations.*

The Chinese have long been proficient mathematicians and believed in finding the best ways to simplify calculations. The abacus is thought to date back to a Chinese mathematician whose name has been lost to history but who is said to have first devised the apparatus more than five thousand years ago. Between this time and the start of the modern scientific age, many thinkers, philosophers and engineers devoted their careers to developing mathematical tools and techniques. They dreamed of labour-saving devices to make calculations faster and more accurate.

The ninth-century Arab mathematician Abu Jafar Muhammad ibn Musa al-Khwarizmi was born in Uzbekistan but spent most of his working life in Baghdad, where he taught at a school, equivalent to a modern-day university, called the House of Wisdom. He is credited with devising the first algorithm, a mathematical technique used to chart the route to solving any given problem. His most famous work was *al-Kitab al-mukhtasar fi hisab al-jabr w'al-muqabala* (*The Compendious Book on Calculation by Completion [or Restoring] and Balancing*) in which he laid some of the foundations of modern algebra. Indeed, the word 'algebra' entered the language from the twelfth-century Latin translation of al-Khwarizmi's book entitled *Liber algebrae et almucabala*.

It is quite possible that al-Khwarizmi experimented with calculating machines in some form but no record of them remains. Others who followed him, especially the Renaissance mathematicians who

---

*One other lasting legacy of Richardson's work is the use of the word 'front' to describe the moving edge of a weather system. This derives from the place in which the meteorologist did his pioneering work.

read the Latin versions of his works, were certainly keen to develop labour-saving calculators. One of them was the great sixteenth-century mathematician Luca Pacioli, who is sometimes referred to as the 'father of accounting'. He designed a primitive calculating machine for speeding up complex financial transactions.

During the seventeenth century the world was graced by some remarkable mathematicians. Newton is perhaps the most famous, but his great rival Gottfried Leibniz was every bit as gifted and he was far better at communicating his ideas. Leibniz once declared: 'It is unworthy of excellent men to lose hours like slaves in the labour of calculations which could safely be relegated to anyone else if machines were used.'[1] He then went on to devise the binary system of calculation which lies at the heart of modern computing and he designed and built a mechanical calculating machine that became the talk of the day within scientific circles and greatly boosted his reputation throughout Europe.

Until the nineteenth century science was guided by the genius and dedication of individuals or small teams who invariably worked without any state backing or any form of support other than that provided by rich patrons or their own private resources. The impact of the Industrial Revolution led politicians to become aware that the state could gain enormously from tapping the intellectual resources of the inventor and the scientist, the engineer and the medic.

One of the best-known examples of this fresh wave of entrepreneurial science came from the work of Charles Babbage, a mathematician who held the Chair of Mathematics at Cambridge University, the Lucasian Professorship, once occupied by Isaac Newton and today held by Stephen Hawking. Babbage was obsessed with calculating machines and he spent a large part of his career perfecting a device he called a 'difference engine'. Babbage was a wealthy man and spent £6000 of his own money in building his device, but he also managed to secure a further £17,000 from the British government after persuading them of the importance of his invention.

This funding came in 1829 when the British Empire was growing. The Napoleonic Wars lay in the past but the British military establishment considered continual expansion and growth to be their first priority. The notion of a machine that could carry out elaborate calculations with great speed was immensely appealing to them, and, because the idea had come from such an important figure with valuable social connections, Babbage gained the support he needed.

Sadly, though, Babbage's machine never performed as well as its creator imagined it could. The truth is that Babbage could have spent £200,000 or £2 million on the machine – it still would not have led to a practical, versatile computer. In retrospect, we can now see that his much-vaunted 'difference engine' bears only a passing resemblance to the simplest modern calculator.

The problem Babbage faced was in some ways similar to the difficulties Luca Pacioli encountered in his time or those that confounded other creative individuals like Gottfried Leibniz. Each of these men had ideas too far ahead of their time: the infrastructure for their inventions was not in place. When Babbage was attempting to build a calculating machine, a mere generation after Volta had created his first battery, the very idea of electricity lay at the fringes of intellectual investigation. Consequently, Babbage was limited to using a mechanical system for his machine, and most of the £23,000 he spent (equivalent to several million pounds today) was used to make the thousands of moving parts that had to be fashioned with previously unimagined precision and made to work in sync with each other. It would take more than a century for technology to catch up with Babbage's dreams of manufacturing a device that could calculate with exceptional speed and accuracy.

Chips with Everything

The crucial component that made the earliest radio, radar, television and computer possible was the valve or vacuum tube. Thomas Edison stumbled upon the electronic principle behind the vacuum tube as early as 1882 but he did not appreciate the significance of what he had observed. It was not until 1904 that a British scientist named John Fleming, who had an intimate understanding of the latest atomic theories, produced the first working model. Two years later, in 1906, Fleming's device was improved upon by the American Lee De Forest, who produced a three-way tube that could accurately and efficiently control the flow of current through a circuit.

Most historical accounts claim that the first electronic computer was a machine called ENIAC (Electronic Numerical Integrator and Computer) which was built at the University of Pennsylvania in 1943 by John Eckert and John Mauchly. They received a military contract worth $500,000 to develop their machine, and by 1945 ENIAC was being used by Los Alamos physicists working on the first atomic bomb. It weighed thirty tons, guzzled electricity and took up 1800 square feet of floor space.

But in fact at least four different groups and individuals should be given equal, if not more, credit for bringing the computer into the world. The American Vannevar Bush (a man who later played a key role in establishing the atomic research centre at Los Alamos in 1942) built a very primitive form of electronic computer in 1930 at MIT. Another was John Atanasoff, a professor of physics and mathematics at Iowa State University, who set out to build a working electronic computer in 1937 and produced a machine that could carry out simple calculations.

Meanwhile, in Britain, the brilliant mathematician Alan Turing led a team that constructed a vast computer called Colossus, which was in operation by 1943 at Bletchley Park and was instrumental in cracking the Enigma code. Turing's work was only declassified many years after the Second World War. In 1954, he faced the possibility of prosecution and disgrace after being exposed as a practising homosexual, and, tragically, he took his own life.

Nevertheless, the man who should be acknowledged as the true father of the computer is the German Konrad Zuse. Born in 1910, Zuse studied engineering. Echoing the pronouncement of Gottfried Leibniz from two and half centuries earlier, that 'It is unworthy of excellent men to lose hours like slaves in the labour of calculations', Zuse was frustrated by the fact that he had to spend so much of his time number-crunching and working through endless mind-numbing calculations, so in the early 1930s he decided to build a computer.

Zuse's first device, the Z1, was a mechanical computer not unlike Babbage's. What was new about the Z1 was that it worked on the binary principle. On-off switches were used to make each decision in the calculating process. By 1940, Zuse had built the Z2, the first fully functioning electronic computer, with which he attracted the attention of the German military, which agreed to fund his work.

During the remaining five years of the war, Zuse perfected his computing machines. They were employed by the Aerodynamic Research Institute which had links with von Braun's group then building the earliest long-range missiles, the V1 and the V2. Ironically, the military became decreasingly interested in Zuse's work as the war progressed, and, when he proposed building a computer using vacuum tubes rather than the electromagnetic relays he had employed in his early models, he was refused funds. The military believed that Germany was so close to winning the war that it had no need for a computer.

But as the war turned against the Germans and the Allies advanced into Europe during the spring of 1945, Zuse became aware that he was in danger and that the Nazis would kill him and destroy his work rather than allow his invention to fall into enemy hands. With his latest machine, the Z4, he escaped from Berlin and fled north to the site of the rocket research installation at Peenemünde. When Peenemünde was overrun by Allied troops, the machine came into the possession of the Americans.

After the war Zuse continued to work at the cutting edge of computer research. In 1958, he constructed one of the first machines to use transistors (the electronic component that super-seded the electronic valve) and he devised the forerunner of modern computer languages: the code used to operate a computer. Although he has been honoured in Germany, Zuse's achievements were overshadowed by the better-known accomplishments of sci-entists who, working quite independently, were nevertheless slower than him to succeed. The Americans Eckert and Mauchly were granted the patent for the computer in 1947.

A huge amount was learned from the earliest computers used towards the end of the war, and because of parallel advances in electronics, which had also been spurred on by military need between 1939 and 1945, the entire science of computing began to advance more rapidly than might have been expected.

Key to the advances in computers' speed was the arrival of more and more sophisticated circuitry. Vacuum tubes are cumbersome and fragile things, and the first ENIAC used no fewer than 18,000 of them. This machine filled a large room, and yet, for all its 18,000 valves, it possessed a computing power roughly equivalent to the rhyme-playing chip in a novelty birthday card.

The leap from ENIAC to the latest computers of the twenty-first century happened thanks to a string of developments in computing, including the creation of programming languages, improved methods of inputting and outputting information, the development of a vast range of software and the enormous and rapid progress in electronics needed to produce increasingly sophisticated hardware.

Perhaps the most crucial advance in computers has been the miniaturisation of electronics. This began during the Second World War with a device called a proximity fuse designed by engineers working at the Royal Aircraft Establishment in Farnborough. The team was led by Samuel Curran who had played a major role in Robert Watson-Watt's team developing radar. After the radar network was established, Curran gathered together a group to work on ways to make the cavity magnetron at the heart of the next generation of radar systems smaller.

One of the traditional ways of attacking ground positions from the air was to drop bombs with timing devices set to explode the bomb after a predetermined period of time. This was limited by the precision timing required and therefore hitting the right target depended upon often fallible human judgement. It seemed that the only way around this problem was to design a device that could guide the bomb automatically.

The Farnborough team came up with a novel way of achieving their objective. They designed a freestanding miniature radar unit, a radio transmitter and receiver that could be fitted into the nose of an artillery shell. This unit could find the target by detecting reflected radio waves in much the same way enemy aircraft or ships are detectable by radar. However, this led to a new problem. How could such a bulky piece of equipment be fitted into the nose cone of a shell? Finding a solution to this problem presented a tremendous challenge, but by focusing their efforts on making vacuum tubes and other electrical components smaller, by 1944 the Farnborough team had succeeded in producing the first electronically guided shells, and they were in use at the front before the war was over.

Although the work of Curran and his team was secret, after the war devices such as the proximity fuse acted as an inspiration to entrepreneurs and civilian designers who grasped the idea that the future of the electronics industry lay with making machines smaller and more powerful. All computers contain large numbers of decision-making circuits, switches that are either 'on' or 'off', and the vacuum tube was crucial because it could turn the current on or off. So, computers employed large numbers of tubes. But because tubes are bulky and cumbersome (averaging several inches in length and girth) and are expensive to manufacture, computer designers quickly concluded that if computers were to advance then tubes had to be replaced with an alternative device that was smaller and cheaper.

The transistor, which replaced the vacuum tube, was developed soon after the Second World War by a team at Bell Laboratories headed by William Shockley who had begun thinking about the theory of such a device during the mid-1930s. A transistor performs the same job as a vacuum tube. It controls the flow of electricity, but, compared with a tube, it is much smaller and cheaper to make. So the transistor was able to revolutionise electronics by making it possible to manufacture circuit boards that were tiny compared with the vast array of tubes and wires that went before them.

With this new invention, the electronics industry took a great leap forward. In 1951, a new industrial complex began to grow near San Francisco in California. Eight thousand acres of land owned by Stanford University were leased to a consortium of pioneering electronics companies led by a company called Arian Associates. By the end of the 1950s, it had become the most important centre of industrial research in America with household names such as Eastman Kodak, General Electric, Lockheed and Hewlett-Packard buying into the place. Within a few years, it had become known as 'Silicon Valley', because of the semiconducting material silicon which lay at the heart of a transistor.

As well as providing industry with the latest innovations in electronics, Silicon Valley became (and remains) a powerhouse for new technologies used by the military. The American government pumped vast amounts of money into developing this centre by subsidising the corporations that moved there, and, by the mid-1960s, more than 70 per cent of products from Silicon Valley were purchased by the US armed forces.

Transistors (or semiconductors) found their place in the commercial world some two years after they were first supplied to the military. The first company to break into this market was Texas Instruments, which correctly saw vast potential in the transistor. To capture public interest they launched the transistor radio in 1954, just as rock 'n' roll was happening and youth culture was booming. Their success led to other companies investing in the miniaturisation of electronic devices and in the same year a Japanese company called Tokyo Tsushin Kogyo Ltd expended huge efforts to break into the American and British markets. At first they made little headway, and, by 1955, they had reached the conclusion that what was holding them back was their unpronounceable name. They quickly changed it to Totsuko, and when this name still failed to spark the imagination of Westerners they rechristened themselves again, this time as Sony.

At the centre of the work done in Silicon Valley lay the transistor and a constant drive to improve electronics, to make circuits smaller, cheaper and more efficient. The spin-offs were the everyday items we have all grown up with – transistor radios, portable TVs, and, more recently, the domestic computer.

The transistor was a great improvement on the vacuum tube, but transistors were still large components compared with the elements of printed circuits used today. As computer designers craved more and more computing power they found they were running up against a size problem. To facilitate greater memory storage and computing power, more transistors were needed. This did not apply to most other devices because they did not require so many transistors to operate and the other components (such as the tube in a TV or a tape mechanism in a cassette player or VCR) were the real size-determining factors.

The solution came in 1961 with the first mass-produced microchip, a revolutionary invention that again sparked a priority dispute. In 1958, two Silicon Valley-based scientists, Jack Kilby at

Texas Instruments and Robert Noyce at Fairchild Semiconductors, independently developed the essential components of a microchip. Both Texas and Fairchild were heavily financed by military contracts, and after a court battle over the rights to the invention which resulted in a sharing of patents to different aspects of the microchip design, the chip was snapped up by the military and employed in a range of innovations at least two years before the public were offered access to it. Since then, the military have contributed almost 60 per cent of all funding for microchip research.

The invention of the microchip came from an entirely new way of looking at electronics. Instead of building a circuit with large components linked by wires, the microchip is a platform of eight different layers of semiconducting silicon and insulators with the components and the wiring imbedded in the layers. This means that elaborate circuits can be made far smaller, and, although the start-up costs of the new technology were high, the expense of producing microchips has fallen rapidly since they first appeared in the commercial marketplace. In 1964, a microchip cost $32 (about $600 in today's terms) to produce; by 1974 it had fallen to $1.27 (equivalent to about $15 now). Today, chips enormously more powerful than their 1974 antecedents cost a fraction of a cent to manufacture.

Two of the earliest big projects to utilise the microchip were the American and Soviet space programmes. Apollo 11, the craft that first took men to the moon, contained no fewer than one million integrated circuits, most of which were microchips in the computer systems of the spacecraft. The lessons learned from the use of microchips alone in the Apollo programme were enough to justify the $50 billion spent on sending men to the moon: they spawned computer and internet industries which together now earn approximately one trillion dollars a year and employ an estimated six million people in the United States alone. But aside from this boon, the development of the microchip, accelerated by the space programme, has altered society profoundly.

Since the first microchip appeared manufacturing techniques

have improved enormously and chips have become smaller and more powerful with each new generation of design. This advance was put into context very early in their evolution when the computer pioneer, and one of the co-founders of Fairchild Semiconductors, Gordon Moore, made the observation that computer advance was exponential and was reflected in the number of transistors per computer. He declared that computer power (or the number of transistors in a machine) would double every eighteen months. The press dubbed this pronouncement 'Moore's Law' and it has been shown to be accurate for more than four decades.

When Moore made his announcement a computer used in a large corporation contained fewer than one hundred transistors; today the average home PC contains in excess of one hundred million. Furthermore, this trend is set to continue until the size of chips becomes constrained by the limits of the physical world. When designers reach the point where they are able to shrink chips to the size of molecules, these chips will become subject to a quantum mechanical rule known as Heisenberg's Uncertainty Principle. This shows that on the atomic level a degree of uncertainty reigns which would actually prevent the chips from functioning as desired. If Moore's Law continues to hold true, this limit is expected to be reached around the year 2020.

The microchip lies at the heart of the computer and some have dubbed it the most important invention of the twentieth century. This is, of course, a familiar mantra, but in this case it might not be such an exaggeration as other claims have been. Many other innovations and inventions could be called the most important of the century, but few people would disagree that the microchip ranks amongst the top five innovations of the last hundred years, and it is one of that rare group of inventions and discoveries that will continue to have a profound influence not just upon technology but upon many wider aspects of society for at least another century to come.

Today, society is enmeshed in the age of the computer. We have become so totally dependent upon these machines and the interactions between them that it is hard to imagine a society functioning without them. Indeed, young people of the early twenty-first century find it almost impossible to visualise a time when there was no such thing as a PC.

One of my favourite after-dinner conversation pieces is to ask those gathered a simple question: 'What do you think would happen if tomorrow every computer in the world stopped working?' A few say, 'Oh, nothing much, there wouldn't be too great a change.' Others, a larger number, suggest that society would revert to how it was in the 1960s, or maybe the 1970s. But the truth is far more dramatic. If every computer in the world ceased to function tomorrow, or even if a substantial proportion of the world's computers shut down, civilisation would end very quickly.

To understand the truth of this statement, we have only to consider how much we rely upon the computer. If there was to be a global collapse of the computer network, defence systems would become inoperative, causing immediate panic throughout military establishments in every developed nation of the world. And, because all communications systems would also be knocked out, no one would be sure if the failure was limited to their own armed forces. Some would imagine that a pre-emptive strike from an enemy could be unleashed at any moment. But deeper than this, the very fabric of society would begin to tear, until, within days, it would lie in shreds.

Think about how we in the West acquire our food, our fuel, the money with which we buy things. The banks would stop functioning. Because the ordering system would have collapsed, no fuel would be delivered to petrol stations and the food in the shops would soon be consumed, never to be replaced. Schools and universities could no longer function; hospitals would be paralysed. Even the supply of electricity and gas to homes, factories and schools, offices and hospitals would be terminated. Soon, civil unrest would be inevitable, violence would erupt, millions would starve, tens of millions more would be caught up in genocide. Within weeks, twenty-first-century man would revert to a Stone Age existence.

One of the things that makes this scenario so frightening is the fact that our dependence upon the computer has come upon us incredibly quickly and without many people seeing the negative aspects. I'm not saying the computer is a bad thing – far from it; I'm a fan of hi-tech and innovation. It is simply that by advancing in this field so far and so fast, for the first time in the history of mankind we have made ourselves incredibly vulnerable just when we think we have reached the pinnacle of civilisation. The danger this complacency represents is every bit as palpable and terrifying as any nuclear nightmare.

Ironically, the biggest stride along this road came at the point I mentioned at the start of this section: the point where the need to know and the need to talk crossed paths; the point in history where computing and communications technology fused. That moment occurred in 1969, the year the internet was created.

Living in Cyberspace

The internet is that rare example of an invention that was created specifically to satisfy a military need. Devised by researchers at the laboratories of DARPA (the Defense Advanced Research Projects Agency), it derived from anxieties and very real fears for the safety of communications and computer networks in the event of a nuclear exchange.

When DARPA (later shortened to ARPA) was established in 1958, at the height of the Cold War, the possibility that civilisation could be snuffed out in a thermonuclear exchange was frighteningly real and military strategists began to devise ways to protect communications and computer-stored information. DARPA was given the responsibility '. . . for the direction or performance of such advanced projects in the field of research and development as the Secretary of Defense shall, from time to time, designate by individual project or by category'.

Concerns were centred upon the idea that if a war began and America was attacked with nuclear weapons, all the communications systems would be disrupted and irreplaceable information destroyed because it was stored at *specific*, and therefore vulnerable, locations. Soon researchers developed the idea of creating a network of computers that could exchange information and continue to function properly even if part of the network was eliminated in a nuclear strike.

ARPA contacted a computer logician at the RAND Corporation named Paul Baran requesting a proposal for a defensive system to serve their needs. Baran submitted a template for what was to become the internet, a system that relied upon a concept known as packet switching, which he described as '... the breaking down of data into "datagrams" or packets that are labeled to indicate the origin and the destination of the information and the forwarding of these packets from one computer to another computer until the information arrives at its final destination computer'.[1] This concept was crucial to the development of a network because if packets are lost at any given point the message can be re-sent by the originator.

Baran is a thinker far ahead of his time and his paradigm for the internet was prophetic in its accuracy. In a recent interview he recalled: 'Around December 1966, I presented a paper at the American Marketing Association called *Marketing in the Year 2000*. I described push-and-pull communications and how we're going to do our shopping via a television set and a virtual department store. If you want to buy a drill, you click on Hardware and that shows Tools and you click on that and go deeper.'[2]

By 1968, ARPA researchers were ready to create their first internet connections when they arranged a link between four separate computers at different locations: the University of California, Los Angeles; Stanford Research Institute; the University of California at Santa Barbara; and the University of Utah.

The experiment was a success but the internet, now used by hundreds of millions of people globally, was rather slow to blossom. The first e-mail was not sent until 1973 and it was to be another eleven years before the number of 'nodes', or users of the network, reached four figures. It was at this point that 'the net', which had been controlled completely by ARPA (and hence the Defense Department), was split into two.

The military wing (MILNET) was appropriated solely for military purposes and researchers in this department developed further the original aims of the programme: the creation of a system that would preserve military information in a nebulous cyberspace away from the destructive forces of nuclear weapons. The other aspect of this technology became known as ARPANET, the civilian wing devoted exclusively to creating a system for use by non-military organisations.

And so things remained for almost another decade. ARPANET was a handy tool for academics and engineers, researchers and computer geeks, and it grew in a modest way. By 1991, a little over six hundred thousand people were connected around the world, but the idea of the internet had not yet seeped into public awareness. The technically proficient and people in the computer industry, as well as the hundreds of thousands employed in universities and civilian research establishments, referred to ARPANET in one of three ways: they talked of electronic mail, the net and the 'information superhighway'. One of these terms, the net, has become part of everyday language; another, electronic mail, has been truncated to e-mail, and the information superhighway has become an anachronism.

The public began to embrace the internet around 1995–6. By this time the world wide web had been created by Tim Berners-Lee, a British scientist working at CERN in Geneva. The world wide web facilitated the transfer of information through the net using hypertext, and this led to the concept of websites, servers, search engines, HTMLs, URLs, Java and all the other elements that go into making the global communications system we have today.

The internet has become ingrained into society like no other technological innovation ever before. It is more important than the telephone or television. It is as much a vehicle for social change as the printing press had been in the fifteenth century. It is bigger than the space industry, bigger even than motorcar production, and it has become as important to civilisation as the aeroplane. The internet extends its tendrils into every aspect of society and it offers an interconnectivity never before imagined.

The internet remains an important tool for the military, but its most visible face is that of a communications tool for everyone with access to a computer and a telephone line. It is becoming increasingly clear that, for better or for worse, communication through cyberspace is also an incomparable instrument for influencing the development of our society. One of the greatest ironies of recent times comes from the fact that, in 2004, a system created to protect the military turned to bite the hand that fed it, for it was through the internet that pictures of tortured prisoners under the care of American soldiers in Abu Ghraib prison, Baghdad, were first revealed to the world.

Epilogue: Some Final Thoughts

There is an ancient proverb that says: 'To prophesy is extremely difficult – especially with regard to the future.' Like many such sayings, this one sounds rather flippant but it is also true. Indeed, it is an aphorism that applies as well to the future of military design and its spin-offs as it does to most other areas of human endeavour.

And yet, while it is impossible to contemplate the exact forms wars will take in the future or how new weapons will affect the way we live in peacetime, one thing is certain: there will be wars and conflict for a very long time to come. And there will continue to be feedback between the military and civilian life which will benefit us all. Furthermore, while it is difficult to visualise any new trends too far ahead, we may consider some that are embryonic and to project a little way into the future.

To begin with, it is safe to say that human beings themselves will become less involved with the physical aspects of war. We are seeing this trend already and it manifests itself in two distinct ways. First, technology has enabled modern armed forces to wage war at a distance. During the two Gulf Wars, Western forces used cruise missiles, satellite-detection systems and unmanned spy planes to find and strike enemy targets. Later, high-altitude bombers were employed. In the future, this strategy will become more widespread. With the use of more advanced surveillance devices and stealth weapons, as well as nanotechnology to produce miniature robot warriors, an enemy will be vulnerable and suffer serious damage without a single casualty from among the attacking force.

Beyond this, most military designers and computer experts believe that in the future a major war may be fought which involves no human soldiers at all. Rather than waging war with physically destructive weapons, it would be far more effective simply to attack an opponent's computer network. This method of waging a war is particularly attractive to maverick leaders of Third World states or international terrorists such as al-Qaeda because they cannot compete with the Western powers in a conventional conflict involving orthodox weapons. The West is totally dependent upon computer networks and any successful attack on this infrastructure would be as devastating to a Western nation's functioning as a nuclear attack.

For this reason, the Western powers are deploying considerable resources to protect themselves against cyber attack, but the fear remains that all networks are in some sense vulnerable. As we grow ever more reliant upon computers to run our lives, the dangers we face become more formidable. The silver lining to this black cloud is that science and technology will reap benefits as ways are found to protect networks better and to combat hackers.

As the West comes under attack from terrorists and pariah states, it will wage a dedicated propaganda war which may bear some comparison with the one waged during the Cold War. Recently, the President of the United States declared his government's intention to push for a Mars mission early this century, and he has tried to gain political points through taking the same stance as John F. Kennedy during the early 1960s when he promised to land a man on the moon before the decade was out. Most scientists are cynical about such statements and probably with very good reason, for the cost of such a mission would be enormous and the technological challenges formidable. But if America is pushed into making powerful gestures and coming through with promises so that it can demonstrate its technological prowess even in the face of attack from 'barbarian terrorists', we will all benefit.

A second sign of physical detachment from the nastiness of war is the way in which Western democracies are increasingly unwilling to risk human lives. The assertion that the West could not sustain a war if too many body bags come home is almost certainly true. In the space of a generation the attitude of the public towards the horrors of war has changed immeasurably.

Hawks would argue that this shift away from risk-taking is a tragedy and that it represents a terrible danger to the West, because such attitudes are not held by the new enemies faced in the twenty-first century. Extremists and terrorists have no qualms about killing innocent children, let alone armed soldiers. Meanwhile, the doves proclaim a more measured and careful approach is a positive thing; they suggest that it implies a healthy reversal in human aggressive impulses, a way forward to a better world.

In this book I have tried to present a cogent argument to support my central contention – that many of the things we take for granted in our everyday lives are spin-offs from weapons development and the waging of war. However, it is important to emphasise that this process is not the only way in which innovation occurs. Inventions have derived from activities and ambitions that have nothing to do with human aggression and the creation of better and deadlier weapons. My hypothesis has been that military need has been the *most influential* and *most varied* influence on technological advance; without it, the modern world would be a very different place.

It is quite possible that many of the technological advances precipitated by war described in this book would have come to us via some other route; that today we would have the car, the aeroplane, keyhole surgery and telecommunications as a result of peaceful pursuits. What is undeniable, though, is that these innovations would not be as advanced as they are if it had not been for war. Without war we might well be travelling in propeller-driven aircraft and steam trains, wondering at the new wonder drug penicillin, while watching a black and white TV transmission of the first men to walk on the surface of the moon – in 2005.

Finally, there is the very matter of the moral thread of the entire hypothesis presented here. I would be the first to admit that this is a thorny issue and that there is probably no definitive position, that there is no perfect stance one may take on this matter. Indeed, it offers a moral quagmire only to be ventured into by the wary. So, allow me, warily, to suggest a path or two through the quagmire and to leave it to you, the reader, to make your own mind up.

It is surely undeniable that human beings are by their nature aggressive. Indeed, most biologists would argue that all living things must possess an instinct for self-preservation. It might therefore be argued that sentient beings, or at least some elements of a sentient race, will always decide that the best form of defence (self-preservation) is to be prepared to attack. Factor in another set of innate drives – such as avarice, jealousy, greed, a need to expand horizons, to possess more than nature provides – and it would seem inevitable that mankind will wage war.

Many would like to believe this is untrue. We all want to live in peace; we all want a safe world for our children. It is simply that such expectations are unrealistic for all and will not be universally attainable for a very long time, if ever. The main reason I say this is because I hold the conviction that human aggression is a necessity. Without aggressive drives, humans cannot achieve. Only dreamers and wishful thinkers could ever suppose that mankind can achieve anything without our inherent aggression. The human drive to succeed and the gift of creativity are inextricably linked at the very deepest level of the 'race psyche' with the need to survive, to prosper, and, in so doing, to do battle if necessary.

If you subscribe to this series of hypotheses then there appears to be no moral dilemma over the argument that good can come from bad and that as well as the pain, death and suffering that have come from war, the technological gains are an acceptable silver lining. If, though, you are convinced that war is not only abhorrent but an aberration, then no benefits, no matter how numerous, can ever begin to redress the balance sheet.

As with almost all moral dilemmas we must accept that in weighing up the positives and negatives of war we cannot take a completely partisan stance, for this is not a matter that may be viewed as being black or white. My personal stance (and I hope this has been conveyed clearly) is that no war is justified because of the technological benefits it might bring. My position is merely to argue that war is inevitable, that it has always been a central theme of human existence and that it will remain so for the foreseeable future. In view of this, we should feel gratified that from this inevitable tragedy of the human condition we may at least salvage some benefits, that from the wreckage may be dragged some small compensation.

Appendix: On the Drawing Board – The Shape of Spin-offs to Come

The advanced research institutions at the cutting edge of military technology play an active and ongoing role in innovating for war. At the same time these agents provide the most fertile source for the new technologies that make such a significant impact on all our lives. What follows is a small sample of the technologies being developed by DARPA (USA) and DERA (Britain), Sandia, NASA and other organisations, technologies that will reach the public domain during the next few decades.

Nanomachines

The potential of nanotechnology was first realised during the 1980s. It is the 'science of small things'. DARPA is currently developing 'killer bees' – insect-size robots for spying and reconnaissance. They are also investigating the design and construction of 'organic air vehicles', small, biodegradable flying spies. The spin-offs from these inventions may include nanomachines for surgery and manufacturing. In the near future nanomachines will be at work in the majority of homes and become as common as the PC or the TV.

Computer simulation

This technology, developed by the military during the Second World War, is now employed by sophisticated machines for battlefield training, and it has been driving the advance of computer games and home entertainment for at least a decade. This technology is used by tacticians and strategists to simulate possible war scenarios and battle plans with far greater ease than ever before. It provides the military with the capacity to carry out virtual manoeuvres and training in a way that is far safer, cheaper and more secure than by using real ships, planes, guns and ammunition.

The software used in computer simulation is extremely sophisticated and requires great processing power to make these virtual exercises more realistic. This drives the development of better computer systems and the evolution of more powerful computers to produce increasingly realistic graphics and sound. Such demands have stepped up with the introduction of 3-D graphics, and even more will be demanded of processors as virtual reality becomes a practical technique. In the not too distant future, soldiers will be donning VR suits and entering a 3-D world of super-realism in which they will train and practise within an almost infinite range of battle scenarios. As a consequence, one day soon, you and I will be able to dress up in specially designed suits bristling with sensors that will allow us to experience virtual worlds.

Such technology will also aid the scientist and the medic, as it is applied, for example, to create more realistic medical scanning and analysers for detecting metal fatigue and ocean-bed pipeline defects.

Battlefield robots

These machines have the potential to change the face of war – robots used to engage enemy robots without ever involving human combatants. Human guidance would come from a control centre behind the lines. Spin-offs from this include robots for domestic civilian use.

Wetware

This is the name given to electronic interfaces between humans and machines. Wetware devices may make it possible for the human mind to control machines directly, making remotely controlled battles a reality. In the civilian world it could be of enormous benefit to people with physical disabilities who could operate machinery and link their brains directly to entertainment systems.

Scramjets

Scramjets are hypersonic aircraft that can travel within the atmosphere at speeds in excess of Mach 6; that is, six times the speed of sound (approaching 4500 mph). They are being developed by the military as advanced fighter and bomber aircraft but they will one day be used as airliners that will travel at around 10,000 mph, allowing them to make the flight from London to Sydney in ninety minutes.

Mars mission

Prompted by a new form of propaganda war, the American government is taking seriously the idea of a manned Mars mission. If it goes ahead, the technological benefits will be as great as, or greater than, the spin-offs from NASA's Apollo missions, many of which have changed our everyday lives radically.

Helmet-mounted cameras

Cameras with integrated displays shown holographically in front of the wearer are being developed for the battlefield but they are also expected to be used widely by firemen, rescue workers and police officers.

Human identification at a distance

Computer software has been developed for the battlefield which allows soldiers and reconnaissance personnel to recognise individuals from a distance. It will soon be adapted for use in banks and to meet a range of security requirements.

Super insulation

A material called aerogel has been developed by NASA as an insulating material for the space shuttle and the International Space Station. Aerogel will soon be adapted for a commercial market where it will be used to insulate homes and vehicles. It will also be used to protect communications cables from radiation and employed as a thermal and radiation insulator for pipelines and electrical cables.

DNA analysers

Developed by NASA, DNA analysers will be installed in spacecraft on future space missions to study the effects of microgravity on the health of astronauts. A DNA analyser separates out cells for study, and in the future this device may be used by doctors on earth to help study tumours, blood and a variety of tissues.

Micro-sample detector

This is a device used to detect microscopic quantities of nuclear, chemical or biochemical materials. As well as being used by anti-terrorism agents, it could also soon be employed by anti-drug agencies and CSI departments. It may also find domestic use as an early detection device for picking up traces of hazardous materials in the home or the workplace.

Voice identification

Computer programmes that can compare a voice to a vast database

for anti-terrorist work may be used by police forces to find missing people or as part of a security system.

Night-vision contact lenses

These devices might be used by special forces to enhance the
visual range of the eyes of field operatives. They also have enormous commercial potential to improve the eyesight of civilians.

Virtual humans

This is another computer programme under development to aid
military training by simulating the movement and mannerisms of
humans. In the future, this software will aid the builders and
designers of advanced VR equipment within the public domain.

Wearable computers

Another sphere in which the military is leading the way with computer advances is in the design of wearable computers. Such
devices have been talked about for at least a decade, but today military organisations such as DARPA and their counterparts in
Britain, France, Russia, Japan and China are all working to produce a combat suit that is smart-wired (with wiring woven into the
fabric of the combat fatigues) and employing a VR visor to project
images such as maps and command instructions.

Such technology will certainly improve the performance of
combat troops, but it will also suffuse quickly into civilian life. It
is possible that before long commuters will be able to follow global
news using glasses to project a 3-D television picture and wearable
computers will become as commonplace as iPods.

Exoskeletons

DARPA are currently developing 'exosuits' that will provide soldiers with added strength and manoeuvrability. In the civilian

world they could prove extremely beneficial to people with physical disabilities.

Translators

Handheld devices to help the soldier on reconnaissance in foreign territory or if captured will be of enormous value to travellers and academics.

References

Introduction: Good From Bad

1. William H. McNeill, *The Pursuit of Power*, p. 331.
2. Leonardo da Vinci, *Codex Atlanticus*, 391r a, Ambrosiana Library, Milan.
3. Sir Solly Zuckerman, *Scientists and War: The Impact of Science on Military and Civil Affairs*, p. x.

Part 1. From the Gods to the Laser Scalpel

The Blood of Men (pp. 13–25)

1. Quoted in John Laffin, *Combat Surgeons*, p. 9.
2. Ibid., p. 10.
3. Ibid., p. 20
4. Ibid.
5. Ibid., p. 22.
6. Ibid., p. 26.
7. Fanny Burney, *The Journals and Letters of Fanny Burney (Madame d'Arblay)*, Letter No. 595 (March to May 1812).
8. Charles Philipps, *La chirurgie de M. Dieffenbach*, Berlin, 1840, p. xiii.
9. Testimony of Dr Thomas H. Killion, Deputy Assistant Secretary of the Army for Research and Technology and Chief Scientist before the House Armed Services Committee Subcommittee on Terrorism, Unconventional Threats and Capabilities Regarding the Department of Defense Science and Technology Policy and Programs, 25 March 2004, published by The House Armed Services Committee, Washington, DC, 2004.

The Fighting Man's Panacea (pp. 26–39)

1. Giuseppe Favaro, *Gabrielle Fallopio Modenese (MDXXIII–MDLXII): studio biographico*, Modena, 1928.
2. Quoted in Richard B. Fisher, *Joseph Lister, 1827–1912*, p. 32.

3. Quoted in Roy Porter, *The Greatest Benefit to Mankind: A Medical History of Humanity from Antiquity to the Present*, p. 372.

The Feminine Touch (pp. 40–5)
1. William Russell, *The Times*, 3 October 1854.

Facing the Horrors of War (pp. 52–7)
1. Quoted in John Laffin, *Combat Surgeons*, p. 118.
2. Quoted in Bruce Shaapiro, 'Lugging the Guts into the Next Room', *Salon Media Circus*, 2004.

Part 2. From the Arrow to Nuclear Power

Guns, Steam and Revolution (pp. 75–86)
1. Charles Hyde, *Technological Change and the British Iron Industry (1700–1870)*.
2. William McNeill, *The Pursuit of Power*, p. 211.
3. Adam Hart-Davis, *Henry Bessemer, Man of Steel*, 'Science and Technology On-line', 1995.
4. Chris Morris, ed., *The Illustrated Journeys of Celia Fiennes*.
5. L.T.R.C. Rolt, *Thomas Newcomen: Prehistory of the Steam Engine*.

Castles and Cannon (pp. 87–92)
1. Leon Battista Alberti, *De Re Aedificatoria*, 1450.
2. Sir William Segar, *Book of Honor and Armes*, 1590.

Guns by the Million (pp. 93–100)
1. 'Tallis's History and Description of the Crystal Palace, and the Exhibition of the World's Industry in 1851', John Tallis & Co., 1852, from Jeffrey A. Auerbach, *The Great Exhibition of 1851*, 1999.

Explosives (pp. 101–8)
1. Quoted in Wang Ling, 'On the Invention and Use of Gunpowder in China', *Isis*, 37, 1947, p. 165.
2. Quoted in Robin Clarke, *The Science of War and Peace*, p. 14.

The Bomb (pp. 109–22)
1. Original in the Hebrew University, Jerusalem, Israel.

Part 3. From Cuneiform to Credit Card

The Written Word (pp. 127–35)
1. Jared Diamond, *Guns, Germs and Steel: A Short History of Everybody for the Last 13,000 Years*, p. 215.
2. 'Warum und wofür?' *Die Wehrmacht*, No. 19, 1939, p. 2.

3. 'Berlin, ein Riesenigel', Hans-Ulrich Arntz, *Das Reich*, 18 March 1945.

Making the World Go Round (pp. 141–8)
1. Quoted in *The History of the Bank of England*, the bank's official web-site at: www.bankofengland.co.uk/history.htm.

Part 4. From the Chariot to the Bullet Train

Horsepower (pp. 163–6)
1. Lynn Townsend White, *Medieval Technology and Social Change.*

All Roads Lead to Rome (pp. 167–72)
1. Thomas Codrington, quoted in Ivan D. Margary, *Roman Roads in Britain.*
2. Quoted in Samuel Smiles, *The Life of Thomas Telford*, p. 3.

The Arrival of the Railway (pp. 173–8)
1. *Quarterly Review*, 21 March 1825, p. 362.
2. Samuel Smiles, *The Life of George Stephenson: Railway Engineer*, p. 205.
3. E. A. Pratt, *The Rise of Rail Power in War and Conquest 1833–1914*, p. 89.
4. Stephen Kern, *The Culture of Time and Space, 1880–1918*, p. 269.
5. Marcel Peschaud, *Politique et Fonctionnement des Transports par Chemin de fer Pendant La Guerre*, Paris, 1926, p. 69.
6. A. J. P. Taylor, *The First World War: An Illustrated History*, p. 20.

The Age of the Car (pp. 179–83)
1. Ransom E. Olds, *Scientific American*, 21 May 1892.

Part 5. From the Balloon to the Space Shuttle

To Fly Like a Bird (pp. 207–17)
1. 'The Wright Brothers Aeroplane' by Orville and Wilbur Wright, *Century Magazine*, September 1908.
2. Ibid.
3. Thomas Edison, quoted in *New York World*, 17 November 1895.
4. Jim Quinn, spokesman for the National Inventor's Hall of Fame, Akron, Ohio, quoted in Amanda Onion, 'From the Laboratory to the Battlefield: Wars and Science Share a Long, Sometimes Conflicted History', ABC News Online, 7 April 2004.

The Rocketeers (pp. 218–39)
1. Quentin Craufurd, *Sketches Chiefly Relating to the History, Religion, Learning and Manners of the Hindoos*, T. Cadell, London, 1790, p. 123.
2. John Harris, *The Recollections of Rifleman Harris as Told to Henry Curling*, p. 98.

3. Konstantin Tsiolkovsky, 'The Investigation of Outer Space by Means of Reaction Apparatus', *Science Survey*, May 1903.
4. Konstantin Tsiolkovsky, 'Investigation of World Spaces', *Aeronautics News*, 1911.
5. Milton Lehman, *This High Man: The Life of Robert H Goddard*, p. 144.
6. Robert Goddard to George Rockwood, 26 December 1916, Ester C. Pendray and G. Edward Goddard, *The Papers of Robert H. Goddard*, Vol. I, p. 121.
7. Quoted in John M. Lonsdon, *The Decision to Go to the Moon*, MIT Press, Cambridge, 1970, p. 128.
8. Federal Aviation Administration (FAA) report published in 2001 based on 1999 statistics. Quoted in 'US Commercial Space Industry Worth $61 billion' by Jeff Foust, *Spaceflight Now*, 8 February 2001.
9. Carl Sagan, *Pale Blue Dot: A Vision of the Human Future in Space*, p. 143; Carl Sagan, London Lecture at Eugene, Oregon, 1970.

Part 6. From the Trireme to the Ocean Liner

Opening Up the World (pp. 245–55)
1. Ethel Wedgwood, ed., *The Memoirs of the Lord of Joinville: A New English Version*, E. P. Dutton and Co., New York, 1906, Part II, p. 96.
2. Quoted in Adrienne Mayor, *Greek Fire, Poison Arrows and Scorpion Bombs: Biological and Chemical Warfare in the Ancient World*, p. 97.

The Naval Arms Race (pp. 271–81)
1. Quoted in Michael Lewis, *The History of the British Navy*, p. 224.
2. Winston Churchill, *The World Crisis*, p. 39.
3. Kaiser Wilhelm II, quoted in the *Daily Telegraph*, London, 8 October 1908.

Beneath the Waves (pp. 282–7)
1. Leonardo da Vinci, *Codex Leicester*, 22b. The Royal Library, Windsor.
2. Quoted in Robin Clarke, *The Science of War and Peace*, p. 110.
3. Ibid., p. 138.

Part 7. From the Tribal Drum to the Internet

Smoke and Mirrors (pp. 293–8)
1. Herodotus, *The History*, Book VI, pp. 105–6.
2. Xenophon, *Hellenica* or *A History of My Times*, Rex Warner, trans, Penguin, Harmondsworth, 1978, Book VIII, 6, pp. 17–18.
3. Francis Dvornik, *Origins of Intelligence Services: The Ancient Near East, Persia, Greece, Rome, Byzantium, the Arab Muslim Empires, China, Muscovy*; B. Peppler, 'Looking Ahead to Intel '97: Security Through Intelligence', *AIPIO News*, 19, 1997, p. 157.

4. *Philosophical Experiments and Observations of the Late Eminent Dr. Robert Hooke*, Royal Society, London, 1726, pp. 142–50 (Originally published 1684).

Messages Through the Lines (pp. 299–310)
1. Quoted in Carolyn Marvin, *When Old Technologies Were New*, p. 101.
2. At a meeting of the Anglo-American Telephone Company on 1 August 1902. Reported in 'Wireless System Not Feared', *Brooklyn Daily Eagle*, 3 August 1902, p. 39.
3. John Meurling and Richard Jeans, 'The Mobile Phone Book: The Invention of The Mobile Phone', *Industry Communications Week International*, London, on behalf of Ericsson Radio Systems, 1994, p. 43.

From Death Rays to Microwaves (pp. 317–24)
1. Tom McArthur and Peter Waddell, *The Secret Life of John Logie Baird*, p. 227.

Numbers (pp. 325–9)
1. Quoted in N. Joachim Lehmann, 'Neue Erfahrungen zur Funktionsfahigkeit von Leibniz' Rechenmaschine', *Studia Leibnitiana*, 1993, 25: 2, pp. 174–88.

Living in Cyberspace (pp. 240–3)
1. Memorandum; *On Distributed Communications: I. Introduction to Distributed Communications Network*, Paul Baran, the United States Air Force under Project RAND-Contract No. AF 49(638)-700, the Defense Documentation Center (DDC). RM-3420-PR, August 1964.
2. Stewart Brand, 'Wired Legends: Founding Father' *Wired*, March 2001.

Bibliography

Adler, Robert, *Medical Firsts: From Hippocrates to the Human Genome*, John Wiley, New York, 2004

Auerbach, Jeffrey A., *The Great Exhibition of 1851: A Nation on Display*, Yale University Press, London, 1999

Aykroyd, W. R., *Three Philosophers: Lavoisier, Priestley and Cavendish*, Heinemann, London, 1935

Berry, Adrian (ed.), *Harrap's Book of Scientific Anecdotes*, Harrap, London, 1989

Black, Edwin, *IMB and the Holocaust: The Strategic Alliance Between Germany and America's Most Powerful Corporation*, Little, Brown, London, 2001

Black, Jeremy, *Warfare: Renaissance to Revolution*, Cambridge University Press, Cambridge, 1996

Bolles, Edmund Blair (ed.), *Galileo's Commandment: An Anthology of Great Science Writing*, Little, Brown, London, 1997

Boorstin, Daniel J., *The Discoverers: A History of Man's Search to Know His World and Himself*, Phoenix Press, London, 1983

Bowen, Daniel, *Encyclopaedia of War Machines*, Octopus Books, London, 1977

Bragg, Melvyn (with Ruth Gardiner), *On Giants' Shoulders: Great Scientists and their Discoveries from Archimedes to DNA*, Hodder & Stoughton, London, 1997

Brennan, Richard P., *Heisenberg Probably Slept Here: The Lives and Ideas of the Great Physicists of the 20th Century*, John Wiley, New York, 1997

Brockman, John (ed.), *The Greatest Inventions of the Past 2,000 Years*, Weidenfeld & Nicolson, London, 2000

Bronowski, Jacob, *Science and Human Values*, Harper & Row, New York, 1956

——, *The Ascent of Man*, BBC Books, London, 1973

Brown, Louis, *A Radar History of the Second World War: Technical and Military Imperatives*, Institute of Physics Publishing, Bristol, 1999

Brownstone, David, and Franck, Irene, *Timelines of War: A Chronology of Warfare from 100,000 BC to the Present*, Little, Brown, London, 1996

Burke, James, *Connections*, Macmillan, London, 1978

Burney, Fanny, *The Journals and Letters of Fanny Burney (Madame d'Arblay)*. Vols I–IX (J. Hemlow et al., eds), Clarendon Press, Oxford, 1975

Carey, John (ed.), *The Faber Book of Science*, Faber & Faber, London, 1995

Cathcart, Brian, *The Fly in the Cathedral: How a Small Group of Cambridge Scientists Won the Race to Split the Atom*, Viking, London, 2004

Chant, Christopher, *The History of the World's Warships*, Regency House Publishing, London, 2000

Cheney, Margaret, *Tesla: Man Out of Time*, Dorset, New York, 1989

Churchill, Winston, *The World Crisis*, abridged and revised edition, London, 1931

Clark, Ronald, *Einstein*, Avon, New York, 1971

Clarke, Arthur C. (ed.), *The Coming of the Space Age*, Meredith Press, New York, 1968

Clarke, Robin, *The Science of War and Peace*, Jonathan Cape, London, 1971

Collins, Michael, *Flying to the Moon and Other Strange Places*, Piccolo, London, 1979

Conn, G., and Turner, H., *The Evolution of the Nuclear Atom*, American Elsevier, New York, 1965

Cook, Chris, and Stevenson, John, *Weapons of War*, Artus Book, London, 1980

Crick, Francis, *What Mad Pursuit: A Personal View of Scientific Discovery*, Penguin, London, 1990

Dampier, W. C., *A History of Science and its Relations with Philosophy and Religion*, Cambridge University Press, Cambridge, 1984

Davies, Kevin, *The Sequence: Inside the Race for the Human Genome*, Weidenfeld & Nicolson, London, 2001

Davis, Nuel Pharr, *Lawrence and Oppenheimer*, Simon & Schuster, New York, 1968

Davis Hanson, Victor, *Why the West Has Won: Carnage and Culture from Salamis to Vietnam*, Faber & Faber, London, 2001

De Souza, Philip, *Seafaring and Civilisation: Maritime Perspectives on World History*, Profile, London, 2001

Diamond, Jared, *Guns, Germs and Steel: A Short History of Everybody for the Last 13,000 Years*, Jonathan Cape, London, 1997

Dornberger, Walter, *V-z*, Viking Press, New York, 1954

Dunnigan, James F., *How to Make War*, HarperCollins, New York, 1993

Dvornik, Francis, *Origins of Intelligence Services: The Ancient Near East, Persia, Greece, Rome, Byzantium, the Arab Muslim Empires, China, Muscovy*, Rutgers University Press, Piscataway, 1974

Dyson, Esther, *Release 2.0: A Design for Living in the Digital Age*, Viking, London, 1997

Eamon, William, *Science and the Secrets of Nature: Books of Secrets in Medieval and Early Modern Culture*, Princeton University Press, Princeton, 1994

Einstein, Albert, *Ideas and Opinions*, Bonanza Books, New York, 1954

Fenichell, Stephen, *Plastic: The Making of a Synthetic Century*, Harper Business, New York, 1996

Ferguson, Charles, and Morris, Charles, *Computer Wars: The Fall of IBM and the Future of Global Technology*, Times Books, New York, 1993

Ferguson, Niall, *The Cash Nexus: Money and Power in the Modern World 1700–2000*, Allen Lane, London, 2001

Fisher, Richard B., *Joseph Lister, 1827–1912*, Macdonald and Jane's, London, 1977

Fontana, David, *The Secret Language of Symbols: A Visual Key to Symbols and Their Meaning*, Pavilion, London, 1993

Friedman, Meyer, and Friedland, Gerald, *Medicine's 10 Greatest Discoveries*, Yale University Press, London, 1998

Fromm, Erich, *The Anatomy of Human Destructiveness*, Pimlico, London, 1997

Galison, Peter, *Einstein's Clock, Poincaré's Maps*, Headline, London, 2003

Gartman, Heinz, *Science as History: The Story of Man's Technological Progress from Steam Engine to Satellite*, Hodder & Stoughton, London, 1961

Gates, Bill, *The Road Ahead*, Viking, New York, 1995

——, *Business @ the Speed of Thought: Using a Digital Nervous System*, Penguin, London, 1999

Gjertsen, Derek, *Science and Philosophy: Past and Present*, Penguin, London, 1989

Goodchild, Peter, *J. Robert Oppenheimer: Shatterer of Worlds*, Houghton Mifflin, Boston, 1980

Gratzer, Walter (ed.), *The Longman Companion to Science*, Longman, London, 1989

Gray, Mike, *Angle of Attack: Harrison Storms and the Race to the Moon*, W. W. Norton, New York, 1992

Greenstein, George, *Portraits of Discovery: Profiles in Scientific Genius*, John Wiley, New York, 1998

Groves, Leslie R., *Now It Can be Told*, Harper & Row, New York, 1962

Harford, James, *Korolev: How One Man Masterminded the Soviet Drive to Beat America to the Moon*, John Wiley, New York, 1997

Harre, Rom, *Great Scientific Experiments: Twenty Experiments that Changed Our View of the World*, Phaidon, Oxford, 1981

Harris, John, *The Recollections of Rifleman Harris as Told to Henry Curling*, Century Publishing, London, 1970

Heisbourg, François, *The Future of Warfare*, Phoenix, London, 1997

Heisenberg, Werner, *Physics and Beyond*, Harper & Row, New York, 1971

Hellman, Hal, *Great Feuds in Technology*, John Wiley, New York, 2004

Henshall, Philip, *The Nuclear Axis: Germany, Japan and the Atom Bomb 1930–1945*, Sutton, Stroud, 2000

Herodotus, *The History*, David Grene, trans, University of Chicago Press, Chicago, 1987

Hersey, John, *Hiroshima*, Penguin, London, 1946

Herzstein, Robert Edwin, *The War That Hitler Won: The Most Infamous Propaganda Campaign in History*, Abacus, London, 1980

Hibbert, Christopher, *The French Revolution*, Penguin, London, 1980

Hogg, O. F. G., *Clubs to Cannon*, Duckworth, London, 1968

Hodges, Andrew, *Turing*, Phoenix, London, 1997

Hoffman, Joseph E., *Leibniz in Paris, 1672–1676. His Growth to Mathematical Maturity*, Cambridge University Press, Cambridge, 1974

Hughes, Jeff, *The Manhattan Project: Big Science and the Atom Bomb*, Icon Books, Thriplow, 2002

Hyde, Charles, *Technological Change and the British Iron Industry (1700–1870)*, Princeton University Press, Princeton, 1977

Irving, David, *The German Atomic Bomb*, Simon & Schuster, London, 1967

Israel, Paul, *Edison: A Life of Invention*, John Wiley, New York, 1998

Josephson, Matthew, *Edison: A Biography*, McGraw-Hill, New York, 1959

Judson, Horace Freeland, *The Eighth Day of Creation: Makers of the Revolution in Biology*, Cold Spring Harbor Laboratory Press, New York, 1996

Kaku, Michio, *Visions: How Science Will Revolutionize the Twenty-First Century*, Oxford University Press, Oxford, 1998

Keegan, John, *The Second World War*, Arrow, London, 1989

——, *A History of Warfare*, Pimlico, London, 1994

Kern, Stephen, *The Culture of Time and Space, 1880–1918*, Weidenfeld & Nicolson, London, 1983

Kevles, Daniel J., *The Physicists: The History of a Scientific Community in Modern America*, Harvard University Press, Cambridge, 1987

Khrushchev, Sergei, *Khrushchev on Khrushchev*, Little, Brown, Boston, 1990

Kline, Morris, *Mathematics and the Physical World*, Thomas Y. Crowell, New York, 1981

Koestler, Arthur, *The Sleepwalkers: A History of Man's Changing Vision of the Universe*, Penguin, London, 1964

Koppes, Clayton, *JPL and the American Space Program*, Yale University Press, New Haven, 1982

Krige, John, and Pestre, Dominique (eds), *Science in the Twentieth Century*, Harwood Academic Publishers, Amsterdam, 1997

Laffin, John, *Combat Surgeons*, Sutton, Stroud, 1999

Lanouette, William (with Szilard, Bela), *Genius in the Shadows: A Biography of Leo Szilard, the Man Behind the Bomb*, Scribner's, New York, 1993

Lattimer, Dick (ed.), *All We Did Was Fly To The Moon*, The Whispering Eagle Press, Gainesville, 1992

Lavoisier, Antoine-Laurent, *The Elements of Chemistry*, Paris, 1789

Lax, Eric, *The Mould in Dr Florey's Coat: The Remarkable True Story of the Penicillin Miracle*, Little, Brown, London, 2004

Lehman, Milton, *This High Man: The Life of Robert H. Goddard*, Farrar, Strauss & Co., New York, 1963

Lewis, Michael, *The History of the British Navy*, Baltimore Press, 1957

Liddell Hart, B. H, *Thoughts on War*, Spellmount, Staplehurst, 1999

Logsdon, John M., *The Decision to Go to the Moon*, MIT Press, Cambridge, 1970

Lomas, Robert, *The Man Who Invented the Twentieth Century: Nikola Tesla, Forgotten Genius of Electricity*, Headline, London, 1999

Lorimer, David (ed.), *The Spirit of Science: From Experiment to Experience*, Floris Books, London, 1998

MacDonald Ross, G., *Leibniz*, Oxford University Press, Oxford, 1984

Machiavelli, Niccolò, *The Art of War*, Da Capo Press, New York, 1965

Macksey, Kenneth, and Woodhouse, William, *The Penguin Encyclopaedia of Modern Warfare*, Viking, London, 1991

Macinnis, Peter, *Rockets: Sulphur, Sputnik and Scramjets*, Allen & Unwin (Australia), St Leonard's, 2003

Margary, Ivan D., *Roman Roads in Britain*, John Baxter, London, 1973

Marvin, Carolyn, *When Old Technologies Were New*, Oxford University Press, Oxford, 1990

Mayor, Adrienne, *Greek Fire, Poison Arrows and Scorpion Bombs: Biological and Chemical Warfare in the Ancient World*, Overlook Press, New York, 2003

McArthur, Tom and Waddell, Peter, *The Secret Life of John Logie Baird*, Orkney Press, Kirkwall, 1990

McKie, Douglas, *Antoine Lavoisier*, Da Capo Press, New York, 1952

McNeill, William, *The Pursuit of Power*, University of Chicago Press, Chicago, 1984

Magee, Bryan, *Popper*, Fontana, London, 1992

Manchester, William, *A World Lit Only By Fire: The Medieval Mind and the Renaissance*, Macmillan, London, 1996

Meadows, Jack (ed.), *The History of Scientific Discovery: The Story of Science Told Through the Lives of Twelve Great Scientists*, Phaidon, Oxford, 1987

Morris, Chris (ed.), *The Illustrated Journeys of Celia Fiennes*, Holt, Reinhart & Winston, 1982

Munson, Kenneth, *Helicopters and Other Rotorcraft Since 1907*, Blandford Press, London, 1968

Naughton, John, *A Brief History of the Future: The Origins of the Internet*, Weidenfeld & Nicolson, London, 1999

Oberg, James E., *Red Star in Orbit*, Random House, New York, 1981

Olby, R.C., et al. (eds), *Companion to the History of Modern Science*, Routledge, London, 1990

Ordway, Frederick I., and Sharpe, Mitchell R., *The Rocket Team*, Thomas Y. Crowell, New York, 1979

Pais, Abraham, *The Genius of Science: A Portrait Gallery of Twentieth-century Physicists*, Oxford University Press, Oxford, 2000

Pendray, Ester C., and Goddard, G. Edward (eds),*The Papers of Robert H. Goddard*, 3 vols, McGraw-Hill, New York, 1970

Pepys, Samuel, *The Shorter Pepys*, Penguin, London, 1985

Porter, Roy, *Man Masters Nature: 25 Centuries of Science*, BBC Books, London, 1987

——, *The Greatest Benefit to Mankind: A Medical History of Humanity from Antiquity to the Present*, HarperCollins, London, 1997

Powers, Thomas, *Heisenberg's War. The Secret History of the German Bomb*, Jonathan Cape, London, 1993

Pratt, E. A,. *The Rise of Rail Power in War and Conquest 1833–1914*, Louisiana State University Press, Baton Rouge, 1965

Read, John, *Explosives*, Penguin, London, 1943

Regis, Ed, *Who Got Einstein's Office?: Eccentricity and Genius at the Princeton Institute for Advanced Study*, Penguin, London, 1989

Rhodes, Richard, *Dark Sun: The Making of the Hydrogen Bomb*, Penguin, London, 1988

——, *The Making of the Atomic Bomb*, Simon & Schuster, New York, 1995

Richter, Jean Paul (ed.), *The Notebooks of Leonardo da Vinci: Compiled and Edited from the Original Manuscripts*, Volumes I and II, Dover, New York, 1970

Ridley, Anthony, *An Illustrated History of Transport*, Heinemann, London, 1976

Rolt, L. T. R. C., *Thomas Newcomen: Prehistory of the Steam Engine*, David & Charles, Newton Abbot, 1963

Rothman, Tony, *Everything's Relative and Other Fables from Science and Technology*, John Wiley, New York, 2003

Sagan, Carl, *Pale Blue Dot: A Vision of the Human Future in Space*, Random House, New York, 1984

Serjeant, Richard, *Louis Pasteur: The Fight Against Disease*, Carousel Books, London, 1973

Shaapiro, Bruce, 'Lugging the Guts into the Next Room', *Salon Media Circus*, 2004.

Shefter, James, *The Race*, Doubleday, New York, 1999

Shepard, Al, and Slayton, Deke, *Moon Shot*, Turner Publishing Inc., Atlanta, 1994

Sherwin, Martin, *A World Destroyed*, Knopf, New York, 1975

Shroyer, Jo Ann, *Secret Mesa: Inside Los Alamos National Laboratory*, John Wiley, New York, 1998

Silver, Brian L., *The Ascent of Science*, Oxford University Press, Oxford, 1998

Simmons, John, *The 100 Most Influential Scientists: A Ranking of the 100 Greatest Scientists Past and Present*, Robinson, London, 1997

Singh, Simon, *The Code Book: A Secret History of Codes and Code-breaking*, Fourth Estate, London, 2000

Smiles, Samuel, *The Life of George Stephenson: Railway Engineer*, Follett, Foster and Company, Columbus, 1859

——, *The Life of Thomas Telford*, World Wide School Library, Seattle, 1997

Sohlman, Ragnar, *The Legacy of Alfred Nobel*, Bodley Head, London, 1983

Spangenburg, Ray and Moser, Diane K., *Wernher von Braun: Space Visionary and Rocket Engineer*, Facts On File, New York, 1995

Stewart, Matthew, *Monturiol's Dream: The Extraordinary Story of the Submarine Inventor Who Wanted to Save the World*, Profile Books, London, 2003

Tarr, László, *The History of the Carriage*, Vision, London, 1969

Taylor, A. J. P., *The First World War: An Illustrated History*, Penguin, Harmondsworth, 1966

Terkel, Studs, *The Good War*, Pantheon, New York, 1984

Tiratsoo, Nick (ed.), *From Blitz to Blair: A New History of Britain since 1939*, Weidenfeld & Nicolson, London, 1997

Tobin, James, *First to Fly*, John Murray, London, 2003

Townsend White, Lynn, *Medieval Technology and Social Change*, Oxford University Press, Oxford, 1966

Turkle, Sherry, *Life on the Screen*, Phoenix, London, 1996

Van Creveld, Martin, *Technology and War: From 2000 B.C. to the Present*, The Free Press, New York, 1991

Van Doren, Charles, *A History of Knowledge: The Pivotal Events, People, and Achievements of World History*, Ballantine Books, New York, 1991

Van Dulken, Stephen, *Inventing the 20th Century: 100 Inventions That Shaped the World*, British Library Press, London, 2000

Warner, Oliver, *The Navy*, Penguin, Harmondsworth, 1968

Warwick, Kevin, *March of the Machines: Why the New Race of Robots Will Rule the World*, Century, London, 1997

Weightman, Gavin, *Signor Marconi's Magic Box: How an Amateur Inventor Defied Scientists and Began the Radio Revolution*, HarperCollins, London, 2003

White, Michael, *Leonardo: The First Scientist*, Little, Brown, London, 2000

——, *Rivals: Conflict as the Fuel of Science*, Secker & Warburg, London, 2001

—— and Gribbin, John, *Einstein: A Life in Science*, Simon & Schuster, London, 1993

Whitfield, Peter, *Landmarks in Western Science: From Prehistory to the Atomic Age*, British Library, 1999

Wilkinson, Frederick, *Guns*, Hamlyn, London, 1970

Wolpert, Lewis, and Richards, Alison, *A Passion For Science: Renowned Scientists Offer Vivid Personal Portraits of their Lives in Science*, Oxford University Press, Oxford, 1988

Wright, Peter, *Tank: The Progress of a Monstrous War Machine*, Faber & Faber, London, 2000

Wrixon, Fred. B., *Harrap's Book of Codes, Ciphers and Secret Languages: A Comprehensive Guide to their History and Use*, Harrap, London, 1989

Xenophon, *Hellenica* or *A History of My Times*, Rex Warner, trans, Penguin, Harmondsworth, 1978

Zuckerman, Sir Solly, *Scientists and War: The Impact of Science on Military and Civil Affairs*, Hamish Hamilton, London, 1966.

Index